Cuatro Ciénegas Basin: An Endangered Hyperdiverse Oasis

Series Editors:

Valeria Souza
Ecology Institute
Universidad Nacional Autónoma de México
Mexico City, Distrito Federal, Mexico

Luis E. Eguiarte
Ecology Institute
Universidad Nacional Autónoma de México
Mexico City, Distrito Federal, Mexico

This book series describes the diversity, ecology, evolution, anthropology, archeology and geology of an unusually diverse site in the desert that is also, paradoxically, one of the most phosphorus deficient sites known. The aim of each book is to promote critical thinking and not only explore the natural history, ecology, evolution and conservation of the oasis, but to also consider various scenarios to unravel the mystery of why this site is the only one of its kind on the planet, how it evolved, and how it has survived for so long.

More information about this series at http://www.springer.com/series/15841

Valeria Souza • Gabriela Olmedo-Álvarez
Luis E. Eguiarte
Editors

Cuatro Ciénegas Ecology, Natural History and Microbiology

Springer

Editors
Valeria Souza
Ecology Institute
Universidad Nacional Autónoma de México
Mexico City, Distrito Federal, Mexico

Luis E. Eguiarte
Ecology Institute
Universidad Nacional Autónoma de México
Mexico City, Distrito Federal, Mexico

Gabriela Olmedo-Álvarez
Laboratorio de Bacteriología Molecular,
Departamento de
Ingeniería Genética
CINVESTAV – Unidad Irapuato
Irapuato, Mexico

ISSN 2523-7284 ISSN 2523-7292 (electronic)
Cuatro Ciénegas Basin: An Endangered Hyperdiverse Oasis
ISBN 978-3-030-06657-4 ISBN 978-3-319-93423-5 (eBook)
https://doi.org/10.1007/978-3-319-93423-5

Softcover re-print of the Hardcover 1st edition 2018

Printed on acid-free paper

This Springer imprint is published by the registered company Springer International Publishing AG part of Springer Nature.
The registered company address is: Gewerbestrasse 11, 6330 Cham, Switzerland

Preface

The Cuatro Ciénegas Basin (CCB), in the northern Mexican state of Coahuila, is an amazing, "magical" place. Upon entering the valley, it is impossible not to be impressed by the high mountains and the blue skies that frame the desert vegetation, punctuated by the Caribbean blue pozas, lakes, and rivers. The town, small, with ordered streets, in the northern edge at the middle of the valley, is rich in history and recognized by the Mexican federal government as *pueblo mágico*. Indeed, a "magic" town because it is especially well preserved and because of its historical and cultural relevance, including different museums, and statues in honor of a native Mexican hero, President Venustiano Carranza, killed in the convulsive years of the Mexican Revolution War. At the main square, there is a beautiful church made from fossil stromatolites. This is a traditional central plaza surrounded by different small hotels and restaurants for the tourist and locals, who are inevitably attracted to the "paletas" store, with the best fruit ice popsicles you can imagine. At the southern border of the town lay side by side the rural middle school and the municipal high school. The later has a techno-agricultural emphasis (CBTA 22), a particularly important player that allowed the scientific team to achieve the results presented in this book.

But besides the touristic value of the place, in the 1960s biologists started to recognize its biological importance. New species of turtles, fishes, aquatic snails, and plants were described in the valley and contiguous areas. The valley was not only beautiful, but had a vast diversity, which was unexpected for a desert, and many of the species were unique to the valley, i.e., were endemic to the area.

In the heroic early stage of the studies, one name is especially important, Professor Wendell L. Minckley. Professor Minkley did not like his first name and preferred to be called by his surname or even better as Minck. In these series of books we will honor him, both its legacy and his preferences. He passed away the 22nd of June 2001. We, Valeria and Luis met him in Cuatro Ciénegas in 1999, in a trip where he and his associates showed us the main aquatic systems and features of the valley (or rather showed them off, as a magnificent treasure!). He explained to us why he believed that this valley represented the remains of a very ancient ocean, still maintained by the springs bringing hot water and different salts from deep inside the Earth. He also commissioned us to take care of the recently declared

"Area Protegida" (protected area) of Cuatro Ciénegas, as he knew or suspected the complicated conservation and administrative problems that the oasis was going to face in the near future. In that trip as well as in other few trips to Arizona, we also discussed why this area was relevant for the study of early life on Earth, a relevant question for the new NASA Astrobiology Institute (NAI) who was on the search of life in extreme environments, believing that those environments might illuminate the dim and fragmented information on the origin of life in the universe. Indeed, CCB is a perfect place for this type of questions: an extreme environment, strongly limited in phosphorus, iron, and nitrogen, where the temperatures can be very high in summer, and also low in winter. Most of all, what made CCB unquestionably unique was the presence and high diversity of stromatolites and microbial mats. Stromatolites are the earliest reliable signals of life on Earth, and nowadays they are only found in extreme environments, where they coexist minimally with animals and other living forms. In a latter visit to Arizona, we discussed with Mink different aspects of CCB's hyper-diversity and we wrote, along with Jim Elser, the grant proposal for NASA that started our scientific studies in Cuatro Ciénegas.

In returning from our first sabbatical in early fall 2000, we stopped at a ranch in Arizona, near the border of Mexico. It was a beautiful "sky island" of diversity in the middle of the desert. Minck invited our children and us to see this place, and probably also to say goodbye. He was coordinating fieldwork for the restoration and conservation of the aquatic environments in his friend's ranch. We had no idea he was very sick, and he died a few months later.

After this initial project of Cuatro Ciénegas with NAI support, more scientists have been involved in its description and study, mostly with Mexican grants. In this first volume of a series, we want to present a first general description of CCB from the point of view of the rich ecosystemic drivers, the microbes.

Not unlike the Viking "Sagas", these book series represent a collection of epic stories, including travels, wars, and misfortunes. In the first chapter of this book, "A Microbial Saga: How to Study an Unexpected Hot Spot of Microbial Biodiversity from Scratch?" by Valeria Souza, Luis E. Eguiarte, James Elser, Michael Travisano, and Gabriela Olmedo-Álvarez, we briefly describe in a chronological order our studies, advances, and mishappenings. It is the story of the slow crawl toward achievements in understanding the microbiota of CCB, from our earliest efforts to the most recent developments. We believe that this first chapter works as a road map for this as well as the next volumes in this series. As Minck told us after the first trip, there are pressing conservation problems at CCB. In this first chapter we explain the strategies used to involve the people in Cuatro Cienegas, to transform them from bystanders into participants of the conservation of their amazing ecological treasure. Local children were involved through art, while the high schoolers participated as collaborators of the scientists, so they not only learned but were empowered, as the owners of their biodiversity, to lead the conservation efforts.

From this general description, in the second chapter, "In the Beginning There was Fire: CCB and the Long History of Life on Earth" by Valeria Souza and Luis E. Eguiarte, we develop an overview of the geological and biogeochemical story of our planet, and how CCB helps us to study and understand this complex history.

In particular, we argue that the colored layers of the microbial mats and stromatolites of CCB recreate the metabolic history of the planet with their different degrees of oxygenation. The deeper layer is the most anoxygenic and is constructed by the most ancient metabolism for which we have a record in the fossil sediments of very old rocks. This recreation of history works similarly to the outdated but intriguing ideas of the early naturalist like Ernst Haeckel, who postulated that ontogeny (the embryological development) recapitulates the phylogeny (evolution) of the animals. We also discussed why CCB is unique and how this story of the past environments and ancient seas of the Earth could be conserved, living and biodiverse, in this peculiar, magical valley.

Moreover, in this first book, we acknowledge that to understand the ecology of a place, we first need to describe its physical environment. In chapter three, we include a detailed analysis of the climate of the CCB. Climate is the first and main determinant of the niche that an organism can occupy. Given the uniqueness of Cuatro Ciénegas, the first stage is to study and describe its biodiversity. Cristina Montiel-González, Francisco Bautista, Carmen Delgado, and Felipe García-Oliva, in their chapter "The Climate of Cuatro Ciénegas Basin: Drivers and Temporal Patterns," describe the original and careful study they conducted using cutting-edge data analyses and report how the climatic variance seems to have increased recently and how global change will affect this already endangered ecosystem, where the average temperature could increase up to 4 °C and the average annual precipitation could decrease 20% given global climatic change. The authors also describe the local patterns of solar radiation, temperature, precipitation, and the concomitant evapotranspiration and aridity index.

After landing the abiotic scenario, we start, in chapter four, with the most abundant and small creatures of this oasis, viruses. "Hyperdiverse Viral Communities in an Oligotrophic Oasis (Cuatro Ciénegas): Marine Affinities and Microgeographic Differentiation" is the work of a large multidisciplinary team (Isa P., Taboada B., Gutiérrez A.L., Chávez, P., del Ángel R.M., Ludert J.E., Espinosa A.C., Eguiarte L. E., Garrido E, López S., Souza V., Arias C. F.) that describes the viral communities found in stromatolites, water, and fishes of the CCB, using metagenomics and bioinformatics approaches. Through infecting and bursting microbes, viruses have an important function in nutrient cycles and regulation of microbial communities. The CCB team observed that indeed viruses are as diverse as the rest of CCB biota, in particular, bacteria. CCB viruses have a unique composition compared with similar viral communities studied in other parts of our planet, with strong marine affinities, as was also suggested from the first bacterial studies. In this chapter, we also explain how this sizable viral community should have a substantial effect on host communities, regulating its sizes and diversity, and also explain, at least in part, the notable diversity of microbes in CCB.

In the next chapter, we show how the studies of the bacteria of Cuatro Ciéngas have relied on the analyses of complete genomes. One of the first microbiological studies published from CCB was the second genome made in Mexico, *Bacillus coahuilensis* (Alcaraz et al. 2008). In chapter five, "How Divergent Is the Cuatro Ciénegas Oasis? Genomics Studies of Microbial Populations and Niche Differentiation" by

Zulema Gomez-Lunar, Mirna Vázquez-Rosas-Landa, Gabriel Yaxal Ponce-Soto, Alejandra Moreno-Letelier, Gabriela Olmedo-Álvarez, Luis E. Eguiarte, and Valeria Souza, we review and compare recent advances in the analyses of *Bacillus* from CCB, in particular, their puzzling auxotrophies (their inability to make specific amino acids), a unique feature within certain *Bacillus* species that suggests that they have evolved a co-dependence on mutualistic strains that can compensate for these amino acids as well as other nutrient deficiencies. In the same chapter, we discuss the comparative genomics of *Vibrio* and other Gammaproteobacteria and the different genes associated with virulence in pathogenic clinical strains. However, at CCB these genes are more related with interactions in the community than with pathogenicity. For example, both the production of antibacterial compounds and its resistance are critical for this community coevolution processes by being able to recognize the "self" in a community that is cohesive and cooperates. The genomic studies in this chapter are opening a door to understanding the very old and divergent bacterial communities in this unique oasis, but they have also become an extremely valuable reference to understand medically important strains. For more than a 100 years, microbiologists have studied mainly medically relevant strains, associating the functions of the different genes and metabolites with particular diseases. Now we can have a broader picture of these functions and study the evolution and adaptations of important bacteria to different hosts.

Both the physical environment and the interaction with other organisms define the element of the niche in bacteria and microorganisms. The study of the niches is also critical to eventually understand the ecology and evolution of bacteria. In chapter six "The Niche at the Edge of Life or the Microbial Ecology (Including Microfungi) of Cuatro Ciénegas: Mutualisms with Locals, Antagonisms Against Foreigners," Patricia Vélez, Laura Espinosa-Asuar, Michael Travisano, Luis E. Eguiarte, and Valeria Souza develop the fundamental ideas for a niche concept for microorganisms by explaining the critical role of microbial interactions in several experiments between bacteria and fungi as well as exploring the role of disturbance. These concepts are critical to understanding the maintenance of the CCB microorganism's antique diversity.

The drivers that can explain diversity are hard to untangle in complex communities. Hence, microbial communities can also be studied in experimental settings, using smaller or larger approaches, i.e., micro- and macrocosms. Different studies involving experimental manipulations both in the field and in the laboratory are described in chapter seven, "Microcosms and Mesocosms: A Way to Test the Resilience of Microbial Communities in Cuatro Ciénegas" by Nguyen E. López-Lozano, Silvia Pajares, Ana E. Escalante, Luis E. Eguiarte, Valeria Souza, and Gabriela Olmedo-Álvarez. The chapter carefully reviews the theoretical basis for the different experimental approaches and describes experiments made with CCB microbes in soil succession in the field, complete biological soil crusts (involving bacteria, fungi, algae, etc.) as well as transplants in a growth chamber and experiments changing the temperature and UV light exposure in the plankton and bacterial mat communities in fish tanks and other *ex situ* long-term experiments currently being conducted in Cinvestav Irapuato using CCB bacterial communities from the sediment. These experiments are critical to eventually disentangle how microbial communities in Cuatro Ciénegas and elsewhere in the world work, not only with field studies, but with actual in situ and ex situ experiments. Sadly, our maintaining replicas of some biological systems is also a precautionary measure to preserve a tiny sample of these endangered world's unique communities.

As mentioned above, one of the first and most studied bacterial genera in Cuatro Ciénegas is *Bacillus*, and now there is a collection with more than 2500 well-characterized cultivated strains. In chapter eight "Single Genus Approach to Understanding Bacterial Diversity, Niche, Distribution, and Genomics: The *Bacillus* in Cuatro Ciénegas," Maria Dolores Rodriguez Torres, Luis David Alcaraz, Valeria Souza, and Gabriela Olmedo-Álvarez describe different adaptations of the genus that accurately correspond to the predictions we had made based on our knowledge of the abiotic conditions present at CCB. This work also describes the different lineages dominating Cuatro Ciénegas. Some of these lineages are unique to CCB and are one of the most important evidence to consider CCB a true "lost world," as these lineages diverged from all other *Bacillus* a long time ago. The authors also discuss different life histories of these *Bacillus,* how they hide, move, and disperse, through their abilities to form spores, to produce biofilms, and to deal with other bacterial antagonistic interactions.

Finally, chapter nine provides a small glimpse of the enormous biotechnological potential of the bacteria of Cuatro Cienegas. The immense and unique bacterial diversity is the result of a large evolutionary history of communities where interactions resulted in co-dependence on other bacteria to survive. The key issue is to distinguish the cheaters from the helpers, as bacteria also possess weapons (like antibiotics) precisely to combat bacterial cheaters and foreigners, even from communities few meters away. Moreover, these bacteria can process and degrade many complex compounds and can survive on very different energy sources besides producing different antibiotics. This biotechnological potential represents a possible avenue to help the local CCB residents to change their economy, always following the Nagoya treaty. In this chapter, we show how we are contributing to develop this biotechnological potential, working with the high school students of CBTA 22, aiming at transforming this sleepy town into a biotech hub. Héctor Arocha-Garza, Hamlet Avilés-Arnaut, Nahui Olin Medina-Chávez, Mario Figueroa, Valeria Souza, and Susana De la Torre-Zavala show us some results in "Actinobacteria and Archaea bioprospecting in Cuatro Ciénegas: a set of open possibilities to save the ecosystem."

Cuatro Ciénegas is not only a beautiful and "magic" place but also a place with rich biological communities that are unique in the planet. This oasis is under strong conservation pressure, as initially narrated to us by Minck. In this book, we describe and compile a collection of chapters describing aspects of the diversity of the tiny witnesses of the amazing CCB geological and biological saga, the surviving microorganisms, their current ecological conditions, as well as essential climatic and geological topics. In the next book of the series, book two, we concentrate on describing biogeochemical and nutrient aspects that make this place so unique. In book three, we write about the diversity and biogeography of animals, while book four will explain the plant's distribution and diversity in the Chihuahuan desert with an emphasis on CCB. The fifth book will address the astrobiological potential of this site comparing it with other sites in the solar system. Near to closing, the sixth book will address the anthropological issues, how the path of humans in this place has changed in history, and why the last 50 years have been negative to the conservation of the aquifer. Last but not least, the seventh book has a series of "love letters" to the ecosystem along with extraordinary photographs. We can only hope this won't become a memorial to this beautiful place.

Finally, this book is important not as an account of a remote place in Mexico, but because this is a place with unparalleled geological and biological treasures that open to humankind the opportunity for amazement, for scientific discoveries, and for understanding Earth's and life history. It is also an example of what scientific collaboration can achieve and brings up the importance that scientific funding can have for scientists in a Country, and this has no precedent in Mexico. CCB made possible the confluence of scientific groups from different research centers and universities, as well as international collaborators, and this books series is evidence of the results obtained so far. This is also a wake-up call for the joint collaboration of citizens, governments, and scientists to seek solutions to the water problem (the elephant in the room) and avenues for the best future possible for CCB.

Mexico City, Mexico — Valeria Souza
Irapuato GTO, Mexico — Gabriela Olmedo-Álvarez
Mexico City, Mexico — Luis E. Eguiarte

Acknowledgements

The research done in this series was mainly funded by WWF-Alianza Carlos Slim, and by SEP-Ciencia Básica CONACYT grant 238245 to both VS and LEE. The books were written during a sabbatical leave of LEE and VS at the University of Minnesota in Peter Tiffin and Michael Travisano laboratories, respectively, both with support of the program PASPA-DGAPA, UNAM.

Contents

Chapter 1
A Microbial Saga: How to Study an Unexpected Hot Spot of Microbial Biodiversity from Scratch?

A Short Introduction to Bacterial Studies in Cuatro Ciénegas

Valeria Souza, Luis E. Eguiarte, James J. Elser, Michael Travisano, and Gabriela Olmedo-Álvarez

Abstract Wendell Minckley had just graduated from college in 1957, and his love was aquatic environments and wildlife management. After his first fortuitous trip to Cuatro Cienegas Basin (CCB) in 1958, his life changed and devoted his career to understand this amazing oasis. In 1997 Minckley, by then a well-established professor at Arizona State University, brought along a new colleague, Jim Elser, a limnologist interested in the basic stoichiometric composition of life. Jim was fascinated by the abundance of stromatolites and microbial mats as the base of the aquatic food web, just as in the early Cambrian sea. What made CCB so special was its extremely low phosphorus and the unbalanced stoichiometry; those were the ancient conditions of the sea in the Precambrian. Jim got the Astrobiology Institute of NASA interested, but they needed Mexican scientists involved as well as experts in microbiology and evolution. Valeria Souza and Luis Eguiarte visited the CCB in 1999,

V. Souza (✉) · L. E. Eguiarte
Departamento de Ecología Evolutiva, Instituto de Ecología, Universidad Nacional Autónoma de México, México City, Mexico
e-mail: souza@unam.mx

J. J. Elser
School of Life Sciences, Arizona State University, Tempe, AZ, USA

Flathead Lake Biological Station, University of Montana, Polson, MT, USA

M. Travisano
College of Biological Sciences, University of Minnesota, St. Paul, MN, USA
e-mail: travisan@umn.edu

G. Olmedo-Álvarez
Laboratorio de Bacteriología Molecular, Departamento de Ingeniería Genética, CINVESTAV – Unidad Irapuato, Irapuato, Mexico

V. Souza et al. (eds.), *Cuatro Ciénegas Ecology, Natural History and Microbiology*, Cuatro Ciénegas Basin: An Endangered Hyperdiverse Oasis,
https://doi.org/10.1007/978-3-319-93423-5_1

and this chapter is about the fascination that they felt at the mysteries behind the enormous microbial biodiversity in the blue-eye ponds (locally called pozas). However, this lost world that had survived eons did not give its secrets lightly, and the hardships to study it are many, so they needed more and more colleagues to join into the task. In the last decade, the now large scientific team has been trying to understand why there are so many species in a place without food, and trying to describe as much as possible and as fast as possible. They all feel this sense of urgency since they are working against the implacable clock of ecosystem degradation. Paradise was disappearing fast given the unsustainable use of water from the deep aquifer to irrigate thirsty alfalfa cultivars in the desert. The decision to save CCB and its amazing biota required to transform a sleepy rural society by translating science to the public and more importantly to give scientific tools to the kids of the owners of the land. Bright-eyed children became conscious about the ecosystem through art, and teenagers are taking actions by discovering that they are the stewards of the biodiversity they are describing in the lab of their high school using molecular tools. There is a revolution occurring in this oasis: science is the tool; kids are the drivers.

How Did All of This Start? Professor Minckley and Cuatro Ciénegas

Wendell ("Minck") Minckley (Fig. 1.1a) was a renowned ichthyologist who was among the first to recognize the amazing diversity of the Cuatro Ciénegas Basin (CCB), doing so very early in his career in 1958. This place in the Chihuahuan desert was a very unlikely site to encounter while starting his master's degree in zoology at the University of Kansas. His first visit stemmed from the mysterious appearance in the zoological collections of an intriguing turtle shell: it was from a species he had never seen before, it was collected from a desert, and it had algae attached. The collector reported that it was an aquatic box turtle. Incredulous, Minck traveled south of the border to see for himself. This was how he first saw the extraordinary oasis of Cuatro Cienegas, the habitat of the endemic *Terraplene coahuila*, the world's only aquatic box turtle. Minck had just graduated from college 1 year before, and his love was aquatic environments and wildlife management. The diversity and the beauty of this Mexican oasis was indeed a paradise for a naturalist, and he spent the rest of his life describing its biodiversity (Minckley 1969, 1992).

Minck was born in 1935 in Kansas, where he went to college, majoring in fisheries and wildlife management and graduating in 1953. He traveled to Cuatro Cienegas dozens of times and was the force behind the declaration of Cuatro Ciénegas as a protected area (APFF, Area de Protección de Flora y Fauna) in 1994. He spent his career describing this unique site and trying to protect it from the mining of the pure gypsum of its dunes. He also believed that the senseless agriculture in the desert was going to destroy it, as the zone was considered "de libre alumbramiento," meaning that landowners were free to exploit surface water. To his dismay, he realized that

Fig. 1.1 Photo collage of researchers involved in Cuatro Ciénegas. (**a**) Picture taken at Jim Elser home in spring 99 where the ASU-UNAM team met for the first time: Valeria Souza to the far left, Jim Elser, Tom Dowling, Minckley is seated. (**b**) Luis Eguiarte, (**c**) Ana Escalante, (**d**) Rene Cerritos, (**e**) Luis David Alacaraz, (**f**) Janet Siefert, (**g**) German Bonilla, (**h**) Gabriela Olmedo, (**i**) Laura Espinosa, (**j**) Michael Travisano, (**k**) Green team from left to right: Eria Rebollar, Maria Rebolleda, Morena Avitia, and Luis Eguiarte. (**l**) Silvia Pajares. Second page of the collage (**m**) Gabiel Yaxal Ponce-Soto, (**n**) Mirna Vázquez, (**o**) Patricia Vélez, (**p**) Jorge Valdivia, (**q**) Nguyen E. López-Lozano, (**r**) Valerie de Anda, (**s**) from left to right, Gabriela Olmedo, Pablo Corcuera, Alejandro Ponce, Maribel Trujano, Valeria Souza, Felipe García-Oliva, and Ariana Hernández, (**t**) students at the molecular lab of CBTA22, (**u**) Nahui Medina, Valeria Souza and Susana de la Torre, (**v**) one of our group photos in February 2017, (**w**) Hector Arocha and the bioprospecting team at CBTA 22

Fig. 1 (continued)

even though Cuatro Cienegas' fishes and turtles, as well as the minerals, were protected by federal ruling, the water was not. That is why Minck decided to shine a spotlight on the ecosystem's importance by bringing a series of colleagues to the basin, hoping that they would initiate studies.

In 1997 Minck, by then a well-established professor at Arizona State University, brought along a new colleague, Jim Elser (Fig. 1.1a red shirt), a limnologist interested in C:N:P (carbon, nitrogen, and phosphorus) ratios, the basic stoichiometric composition of life. Elser developed a fascination with the streams and ponds of Cuatro Cienegas and especially its microbial life as manifested in the "living stromatolites" of the basin and how they were affected by CCB's geochemical condition. Jim saw an opportunity to fund work at CCB via NASA's astrobiology program and developed a research component for a large proposal that the university submitted. This was a compelling idea, given astrobiology's dual focus on life on early Earth and on possibilities for life beyond Earth. Astrobiologists have proposed that the low supplies of P on the Earth's Archean ocean and during all the Precambrian were important in shaping early evolution. It was also speculated that Mars' initial ocean was also low in P, as were our early seas. Hence, CCB was the perfect analog for both Earth and Mars' early oceans: sites rich in sulfur and magnesium but that are very limited in P. CCB ecosystems are low in P and strongly imbalanced in N:P stoichiometry, meaning that P limitation is likely severe. Furthermore, CCB harbors more diversity of fresh water stromatolites and microbial mats that anywhere in the world, making it an even more attractive site to study as an early life analog.

With this information, Elser and his collaborators at the Arizona State University developed a component for the grant proposal that aimed to establish ASU as a core team member of NASA'S Astrobiology Institute. The idea was to uncover how life copes with extremely nutrient-deficient conditions. However, the team consisted of a geochemist expert in early Earth, two fish experts, a limnologist, and an expert in mathematical models. Feedback on the proposal from NASA requested that they increase the team's expertise in microbiology and evolutionary biology and that they engage Mexican scientists. Based on a colleague's recommendation, Jim found this combination of skills and nationality in the National Autonomous University of México (UNAM): Luis E. Eguiarte (Fig. 1.1b), interested in population genetics and molecular evolution and Valeria Souza (in blue at Fig. 1.1a), an expert in evolutionary ecology of bacteria. Valeria and Luis happen to be a couple, and in spring 1999 the complete Eguiarte-Souza family (that included 6-year-old María and 11-year old-Felipe) joined Minck in the small sleepy town of Cuatro Ciénegas. He arranged for them to visit many exciting sites, explaining them the particularities of each poza. Minck made a particularly bold statement in the beautiful spring of La Becerra: "Do you see these miniature snails in my hand? I just scooped them from the spring-head, but their direct ancestors were eating sulfur bacteria in hydrothermal vents 220 million years ago in the bottom of the ancient Pacific (Panthalassa). Now they are here, trapped in a desert oasis, eating the same bacteria that they used to eat. It is your job to figure out how these snails survived for so long along with their ecosystem, and only *you*, as Mexicans, can save them from the extinction caused by humans." Minckley's intuition about the marine origins of these snails and their habitats was right, even if he could not prove it at the time since he did not have the molecular tools that the new Mexican team members had been using since their PhD studies. Funding from NASA came soon thereafter to support this new team with the Astrobiology Institute. A new era of Cuatro Cienegas science had begun.

Is Cuatro Ciénegas Really an Ancient Sea?

After collecting permits granted by the Mexican government, the ASU-UNAM team made the first field trip with astrobiology funding to CCB. They brought with them many ideas and numerous Petri dishes with all the different media they could get. To their surprise, microbes were not only abundant but very colorful and diverse in colony morphology. However, when we tried to get axenic (pure) cultures, diversity became very low. We also tried to isolate environmental DNA, inspired by a paper we read about a sulfur cave that made use of such novel technique. However, DNA isolation with standard methods did not work—the high magnesium concentrations in the water as well as a large variety of molecular products of bacteria inhibited the PCR reactions that are needed to do metagenomics. From the very beginning, it was clear for all the team that this place was odd and hard to explain and even harder to study than what we predicted. We had a plethora of questions: Why was this place so diverse, not only in terms of plants, invertebrates, and fishes but also in microbes, microbial mats, and stromatolites? Why was there water in this valley and not in the other valleys around it? Why did the basin have this "butterfly" shape, with high sierras that looked like they had been sculpted by giants? As tectonic processes were evident to the bare eye, could these imposing mountains be related to Minckley's hypothesis of the ancient sea?

While several students and technicians tried to tackle the DNA problem (Ana Escalante Fig. 1.1c and Laura Espinosa, Fig. 1.1i), Valeria spent the first 2 years of the project rethinking everything, analyzing maps at different scales with the mountains, and studying what little was known about what seemed to be a very complex fault system. She was trying to disentangle the water, its origins, and its movements. At the same time, even though we were still struggling to obtain environmental DNA, we began to cultivate many lineages of bacteria, obtaining their DNA, and sending it to be sequenced. When the results came back, we discovered that, among the relatively few sequences available in the GenBank database in those days, the only sequences that they matched were those from marine bacteria. Valeria did not doubt an instant about the meaning of these odd data: Minck was right, the ocean creatures survived in CCB. However, many disbelieved the new data. At the same time, Luis and Valeria's long-time friend, Michael Travisano (then at the University of Houston (Fig. 1.1j), came in 2001 to help them think about how to design studies to explain the diversity of microbes in the basin.

An October 2002 field trip looked like total madness: at the same time that Jim and his ASU team began experiments with snails feeding on oncolites (a type of rounded stromatolites found in Rio Mezquites, Fig. 1.2b; see Chap. 5 in Book 2), the UNAM team was developing the first in situ DNA isolation scheme. Their hypothesis was that, perhaps, DNA degraded too fast and did not survive the trip to Mexico City in liquid nitrogen. If so, then we needed to clean it on-site. In this case, this was in the hotel's little kitchen. For Ana's work, we needed to get water samples from different sites, filter them, and extract their DNA. At the same time, a film crew from National Geographic was filming the field trip, as speleologists and hydrologists from the Italian caving group La Venta were going to explore, along with the UNAM team, the source of the valley's water and its mysteries. At this time, Valeria had a "eureka moment": the crazy idea of testing for the lost ocean (Minck had sadly passed away in 2001) via deep samples obtained by the La Venta speleologists while the film crew filmed. It was a good idea, as we were able to take samples from different deep springs and from an abandoned mine shaft 800 m below the surface and then gave them to the UNAM team for DNA extraction. To make things even more interesting, Felipe Eguiarte (Luis and Valeria's son) had organized a field trip with his middle school to see how science developed in this oasis; 30 kids and 3 teachers were dazzled forever by the beauty and the mysteries of CCB.

New students continued to join the project. Ana Escalante (Fig. 1.1c) was starting her PhD dissertation, aiming to explain the bacterial diversity of the Churince (Fig. 1.2a) water system that dominates the western portion of the butterfly-shaped basin of CCB (Fig. 1.2). Her work focused on the *Pseudomonas* genus, a lineage of bacteria that we could easily cultivate. Since she wanted to understand the community context where that lineage was living, Ana went to the University of Idaho, in Moscow, Idaho, to the laboratory of Larry Forney, with freshly extracted DNA from the ecosystem to carry out fingerprinting of the microbial community.

Following on Minckley's Steps: To Understand and to Protect

In that same trip of October 2002, Valeria was trying to convey Minckley's hypothesis of the ancient sea to Italians who did not speak either good Spanish or English (and Valeria does not speak any Italian). Concerned town people approached the UNAM team because turtles were dying in the most southwestern poza, Nueva Atalaya (the closest to the Hundido valley), and the river that used to cross the town was dry. Their concern was that recent developments associated with giant dairy companies in the neighboring valleys (in Ocampo-Calaveras to the north and in the Hundido to the south) were apparently depleting the aquifer. Concerns were strong, as water demands would only increase in the future. These companies had planned to drill 200 deep wells in the desert in the valley of El Hundido in 2002–2003. Each of those wells would be able to irrigate ca. 70 ha of alfalfa, a cultivar that needs a very large amount of water and needed to be wet even in the 50 °C summer temperatures. By then, Valeria, who had been drawing fault systems in maps and reading all she could find on the geological, geophysical, and hydrological papers from the area, knew that such an agriculture project would be a death sentence for the diversity of CCB.

Fig. 1.2 Mexico and in the state of Coahuila, south of Texas, the Cuatro Cienegas Basin. (**a**) Churince Laguna Intermedia spring 2016, (**b**) Oncolite at rio Mezquites, (**c**) Pozas Azules stromatolites, (**d**) one of the ponds at Pozas Rojas

Since we were sampling for Ana's innocent dissertation, it was easy to extend the sampling to water from the deep wells that irrigated alfalfa in both of the neighboring valleys (Souza et al. 2006). A new student, Laura Espinosa-Asuar (Fig. 1.1i), was recruited to obtain clones from the environmental DNA and to complement Ana's "community signature" study. The first community data that Ana got at the University of Idaho via TRFL analysis showed that, indeed, the communities not only were related to marine communities but that the three valleys (CCB, Ocampo-Calaveras, and Hundido) shared many microbial members (and none with other environments). Hence, they were connected. Laura's 16 sDNA clone libraries also showed that more than half of bacteria from the different sites had a marine origin and that many of the sequences (in particular, those obtained from the deep sample from the mine shaft) were associated to hydrothermal vents, just as Minck had predicted (Souza et al. 2006).

The description of the shared communities in the valleys, and of the initial evidence for the marine origins of the basin's microbes, gave the UNAM team evidence-based "ammunition" to convince the federal and local government to restrict agricultural development to save this lost sea. Ana kept on working on her dissertation, publishing a first paper describing community heterogeneity in the Churince (Fig. 1.2a) hydrological system (Escalante et al. 2008) and then a second one describing a new species of *Pseudomonas* from a sampling site in the Churince system that, sadly, no longer existed by the time of its publication (Escalante et al. 2009). Her sample originally collected in Churince's terminal lake, Laguna Grande of Churince, had dried completely at the end of the decade.

By then, the UNAM team received the first of a series of grants from CONACyT, the science council in Mexico, and from UNAM. With this funding, another study followed the interesting temporal variation of *Pseudomonas* spp. from summer vs. winter in the Laguna Grande of Churince. This work was also published as a memorial to this amazing, but now dry, site (Rodríguez-Verdugo et al. 2012). To deepen into the understanding of these bacteria, UNAM enlarged its team by adding Professor Gloria Soberón, an expert on *Pseudomonas* secondary metabolism, contributing another PhD dissertation (Toribio et al. 2011) that described for the first time the unique biotechnological potential of the site.

During that 2002 trip, Jim's ASU group continued its cunning experiments of snails perching on round stromatolites from the Mezquites river (Fig. 1.2b), subjecting them to different concentrations of phosphorus. While the ASU team determined how C:N:P ratios responded to P enrichment via changes in photosynthesis and calcification, Ana, Valeria, and Luis (along with other colleagues from ASU) worked on describing the changes in the microbial community (Elser et al. 2005a, b). In their experiments, we showed that snail performance was stimulated by small amounts of P fertilization (likely because their low-quality, low-P food got a little richer in P), snails got smaller and suffered mortality as P increased further, a surprising observation. From these findings, Jim and his group concluded that CCB's snails seem to live on a "stoichiometric knife edge" of optimal nutrients, an inference with potential implications for understanding the evolution of Earth's first higher trophic levels during the events preceding and following the Cambrian explosion (Elser et al. 2006).

During this period, a former student of Minck's, Thomas Dowling (Fig. 1.1a, cream shirt), had his own Ph. D. student, Evan Carson, looking at the fish genetic structure (how populations differentiate genetically in space) and hybrid zones in the Churince system. To his dismay, the number of fishes became fewer and fewer with time (Carson and Dowling 2006; Carson et al. 2012, 2013, 2015). Another team was evaluating snail diversity, finding that there were micro-endemisms at each site. During that time, another astrobiologist, in this case from Rice University in Houston, Texas, Janet Siefert (Fig. 1.1f), got involved in the research bringing more people from NASA to study the characteristics and diversity as well as the beauty of the site.

In 2005, Forest Rowher and Mya Breitbart, virus experts from San Diego State University, were invited to enlarge the knowledge of this site, since cultivating virus (one of Mike Travisano projects) had proved impossible. The reason for Mike's frustration was now evident: based on the metagenomics data Rowher and Breitbart completed, the diversity was enormous, and most viruses were divergent and of marine origin (Desnues et al. 2008). Thus, a Petri dish with a culture was a very unlikely bet to fish for dormant viruses in those cultures. The San Diego team also did the first metagenomics studies of CCB (Breitbart et al. 2009; Nitti et al. 2012), describing the microbial diversity of the stromatolites in Pozas Azules (Fig. 1.2c) and Rio Mezquites.

On the other hand, selective media did recover multi-antibiotic-resistant bacteria and, in particular, we learned that Marine Medium enriched for heat-resistant *Bacillus*. In late spring 2003, Rene Cerritos, another UNAM student (Fig. 1.1d), started to work with the thermoresistant cultivated strains and discovered an enormous diversity, especially in the much endangered big lagoon (Laguna Grande) of Churince. He described the physicochemical composition of water along the Churince water system to correlate that diversity to the environment. The system (Fig. 1.2a) comprised a springhead (Manantial), a river (Río), an intermediate lagoon (Laguna Intermedia), and the large desiccation lagoon (Laguna Grande). It sounded easy. So, Valeria and Luis convinced their friends Marco Araujo, Pablo Vinuesa, and Mike Travisano that since Churince could soon disappear, we needed to know its hydrology. Marco and his friend Chris obtained a new instrument (a sophisticated type of hydrolab) that could profile many parameters from a boat. However, as the water was lower than expected and the river that connected the system had gotten smaller, the small row boat that was supposed to make it float was too wide for the small river, and the boat, with its heavy instrument, could not easily float in the shallow pools. Eventually, the boat was carried by five people while the machine was dragged for hundreds of meters by another resilient team. This led to coining of a new CCB motto: "maximum effort for the smallest result." Nevertheless, the analysis showed that there were many different small springs in Laguna Intermedia and that this lagoon was not only in the middle of the system (that is what Intermedia means) but that its physicochemical conditions were a gradient between the saltier, more marine Laguna Grande and the freshwater-dominated springhead. This helped not only Rene Cerritos in improving his sampling of heat-tolerant bacteria (*Actinobacteria* and *Bacillus* spp) but also the Mexican team in better understanding the sources of water in Churince (Cerritos et al. 2008, 2010).

The "maximum effort" motto end up including both field and lab work. In the lab, the hardest was extracting clean DNA for the metagenomic samples. If extracting

DNA from water was very tricky, as the first studies with fingerprinting showed, then imagine what it was like to do it and clean it from microbial mats or stromatolites or the soil. The problem was that the PCR reactions involving DNA from CCB were consistently inhibited, at least in part due to high Mg and lots of polysaccharides but also due to small mysterious peptides that interacted with Taq polymerase. We like to imagine these inhibitors as an ancient defense mechanism against strangers. It took 2 years to polish the methods. However, it was worth the effort, as now we could extract DNA from the soil, sediment, and stromatolites and keep on describing the site. On most of the field work adventures were constant, from getting stuck in the semi-wet sediment while walking to sample the Laguna Grande when it was starting to disappear, to countless off-road trips to get to the pozas in the middle of the basin where vehicles got stuck in the mud or had flat tires due to the mesquite spines. We also encountered some angry ranchers with rifles in hand who had to be persuaded that sampling was legal and that everything was done according to the law and approved by federal, state, and local authorities (there is a federal collecting permit that UNAM received and renewed each year).

With the philosophy that it takes a village to raise consciousness, CCB became a magnet for friends and new scientists, including an amazing team of Mexican scientists from CINVESTAV-Irapuato: Professors Luis Herrera-Estrella and Gabriela Olmedo-Álvarez (Fig. 1.1h), with her PhD student, Luis David Alcaraz (Fig. 1.1e). In 2006 Cinvestav-Langebio, the national laboratory of genomics and biodiversity in México directed by Luis Herrera-Estrella, was new and wanted to test the first 454 DNA sequencers. So, he proposed to sequence the first bacterial genome of CCB. We chose an endemic *Bacillus* from Rene Cerritos' collections, one obtained from the now extinct Laguna Grande in Churince (Fig. 1.2a: Cerritos et al. 2008). The CINVESTAV scientists wanted to see the site where the bacteria used to live; Gabriela was enthralled by all the possibilities and forgot a little bit that she had become a hard-core molecular biologist and returned to her field biologist roots.

The genome of *B. coahuilensis* also showed marine affinities as well as unique adaptations to the site's low phosphorus content, such as making membrane sulpholipids instead of phospholipids and using a solar antenna to survive if nothing else was available (Alcaraz et al. 2008; Gómez-Lunar et al. 2016). Another genome from a *Bacillus* spp. from the same site showed another adaptation, such as the use of different sources of phosphorus, and had an even older marine ancestry (Alcaraz et al. 2010). Rene Cerritos meanwhile described an amazing diversity of *Actinobacteria* and *Bacillus* in Churince (Cerritos et al. 2010), many of them marine-derived and unique and, interestingly, showing a mostly clonal structure.

These unexpected data moved Valeria, Luis, Janet, and Jim to write a paper on the potential role of P in bacterial sex and speciation (Souza et al. 2008), arriving to the conclusion that "no food means no sex which means lots of local diversification," arguing that this was the reason behind CCB's extraordinary microbial biodiversity (see also Souza et al. 2012a, b). This model was supported in a very nice study on the genetic population structure in *Exiguobacterium*, a mostly extremophile genera with several endemic species in CCB (Rebollar et al. 2012) and confirmed in

Pseudomona and other *Bacillus* (Avitia et al. 2015) (Fig. 1.1k). Moreover, when dating with molecular clock many of the lineages of *Bacillus*, they showed to be not only marine but extremely ancient, some of them dating back to the Precambrian (Moreno-Letelier et al. 2011, 2012).

Since the arrival of Mike Travisano to CCB in 2001, we were thinking about how to do experimental microbial ecology to test various hypotheses. It was hard to understand why there was so much diversity in a place without food by focusing only on population genetics surveys or on the "total picture" that a community metagenome study could provide. We needed something in between, field mesocosms or experiments in the lab. With Gabriela Olmedo, we designed studies with artificial substrata that became colonized by the sediment communities of the Churince and using them to study the interactions between its members, particularly the *Bacillus* spp. The studies showed that competition among its members through antagonistic interactions was an important mechanism by which the communities are formed (Pérez-Gutiérrez et al. 2012). Meanwhile, German Bonilla (Fig. 1.1g) and Mariana Peimbert joined the team to extract high-quality DNA from mats from two contrasting sites and carried out metagenomic sequencing with methods that allow direct sequencing of the DNA with high coverage (Bonilla-Rosso et al. 2012, 2013; Peimbert et al. 2012). The data showed that each community was unique, and all of them were metabolically "complete," that is, their functional diversity was very large despite having different microbial compositions.

To get into more complex settings, Nguyen Esmeralda López Lozano (Fig. 1.1q), a new student in 2005, did succession experiments in soil that followed the microbial diversity and the nitrogen cycle with molecular markers. She chose Churince's soil with two sites, one close to the river in a "wet zone" with microbial mats and another in the dry part in the former location of Laguna Grande. She blasted gamma rays in 64 mesh bags of 1 kg of soil taken from each site (32 each); then, the bags where "re-seeded" in the holes from where the soil was taken. Succession was followed for a year, sampling every 3 months (López-Lozano et al. 2012, 2013). This experiment involved geochemical knowledge, so we invited to the team professor Felipe Garcia Oliva (Fig. 1.1s, black tee shirt to the right), also from UNAM. Despite the harshness of the environment and its technical difficulties, he was hooked and brought students and ideas who started to unmask the mystery of the soil's C-, N-, and, of course, P-nutrient cycling along with the vegetation physiology (Perroni et al. 2014, 2015; Tapia-Torres et al. 2015, 2016). These included changes due to agriculture in the desert (Hernández-Becerra et al. 2016). His team observed that CCB's soil microbiome was amazingly well-tuned to its low-nutrient regime, much like the "knife edge" equilibrium that Elser had found in the small snails. In fact, if fertilizers were added, the soil microbiome changed as opportunistic microbes broke the fragile N:P equilibrium that allowed plants to grow.

Water system mesocosms went through several iterations. With Mike Travisano, we purchased many bathroom tiles, cleaned them, and placed them in many pozas to allow them to grow a microbial community on top. Demonstrating our "maximum work, minimum results" motto, they did not work: they were too heavy and sank into the soft sediments. The second iteration was the good one. We designed acrylic trays for microscope glass slides and used a pole to fix them at the bottom of the water column. We placed three poles, with three trays each, in two pozas in the Pozas Azules at the Pronatura Ranch. The colonization of the glass slides was observed for days and weeks, and the poles were left for 2 years to get fully biodiverse and "fluffy." In Summer of 2009, Jess Corman started to study the C:N:P in the stromatolites in CCB and the effects of calcium carbonate deposition on P availability in stromatolites (Corman et al. 2016). In addition, she and ASU PhD student Jorge Ramos performed a broad survey of the hydrology and composition of all the wetlands and pozas for comparison with old records of the sites. At the same time, Silvia Pajares (Fig. 1.11), a new postdoc in the UNAM team, designed a test to study not only the glass slides' biodiversity but also the effect that different environmental conditions would have on the biodiversity. This involved taking the slides and placing them in 15 fish tanks (five treatments, three replicates) with water from the pozas (40 L each) to test several global change scenarios: high UV, no UV (as control), higher temperature (40 °C), control (30 °C, that is the average temperature of these pozas), and a fluctuating environment without temperature control simulating a shallow poza. Imagine now transporting 600 L of water (15 × 40 L), by hand, through the desert for miles. We needed an army. So, Luis and Valeria invited several students and their kids' friends to join them at CCB. We all had a lot of fun setting up the mesocosms. The best part was that the mesocosms were not only Silvia Pajares' job but their maintenance and care also became a science project of the local high school (Pajares et al. 2012, 2013, 2015).

How to Save a Paradise: Involve the Kids

The "maximum effort for the smallest result" motto applied for Silvia's mesocosm experiment, but it would be impossible logistically to do a year-long experiment without the local students. How we got that valuable and trained help?

In the early 2007, it seemed that UNAM team had won a rough 4-year political battle with the media on their side. However, a few angry locals claimed that science did not belong in their town and that scientists were infecting the town with the dangerous "virus of conservation." This "infection" increased on February of 2007 as the UNAM team organized a "VIP trip" to CCB with more than 80 high profile guests that included all the ASU-UNAM team, international press, members of the National Academy of Sciences (USA), opinion leaders, politicians, friendly scientists, and corporate representatives. The logistics included nearly 20 vans (funded by the Coahuila government) to move along the desert's tricky roads. Everybody was amazed by the beauty, impressed by the scientific significance, and pained by the possible loss of this desert oasis.

Water wars are not only fought in the halls of government; it must also be an everyday task of the locals. So, in the midst of the media battle (2002–2006), the local teachers decided to start their own revolution, a transformation of society through education with the help of the Mexican researchers. The UNAM team had been, since 2003, giving town hall talks about their research. We did so to start to spread this "thirst for knowledge" to the teachers and the students from the local high school, CBTA22 (Centro Bachilletaro Tecnologico y Agropecuario, the only high school in the county). The idea was that the description and discovery of new biodiversity is a community task, even more so is the conservation of this extraordinary ecosystem.

Since 2004, during every field trip, the UNAM team went to the CBTA22 high school to give a talk about their ongoing and planned studies and discoveries that had emerged. Few people beside the students attended at the beginning but, trip after trip, we involved teachers and students in discussing results together. The discussions became more popular. From those interactions, we found out that, with simple replicable experiments, the kids could diagnose the state of the aquifer using several biological markers that UNAM team had developed for them. The approach involved "seeding" microscope glass slides in the water for a week and retrieving them to look at the microscope. The approach was very visual—the more diverse in shapes and colors, the better since they reflected healthier ecosystems. The sleepy town society was awakening, as the teenagers pushed to learn more and, at the same time and reflecting a more informed population, tourism started to also change from predatory "cheap beer-loud music-leave the trash" type to a more conscientious eco-sustainable type.

This work culminated in 2009, when an extraordinary team of 10 students and their teachers worked on the fish tanks that Silvia had set up to record all the changes during a complete year. They saw the differences in the communities along time using microscopes (just as they did before to see the health of the pozas) and determined pH, temperature, and color changes. In summer 2010, the students went to Mexico City to the molecular lab at UNAM to learn how to isolate DNA that was then sequenced. The UNAM team then shared the results with the kids, and now they knew the name of the microbes that they were taking care of.

Churince: The Most-Studied Little Place on Earth

In 2009, the WWF-Fundación Carlos Slim approached UNAM team to see how they could help. Churince was special—it was the most divergent and unique place even on the CCB scale of uniqueness. It was mostly Jurassic marine sediment, with a strong magmatic influence from the crustal anomaly that uplifted the Sierra San Marcos at the center of the butterfly-shaped basin. In early 2010, while Luis and Valeria were on sabbatical at the University of California at Irvine, they got the news that not only Carlos Slim wanted to buy the land for conservation but that it was giving them a grant for 5 years to study "everything." Between 2003 and 2010, the CCB team had several grants from Mexican CONACyT and UNAM to support their research, but this was a very welcome addition and a break from grant writing (but not from grant reports, which were required every 2 months or so).

The new funding from Carlos Slim coincided, luckily, with another NASA grant led by Elser that began in Fall 2010. The NASA grant was named "Follow the Elements" and supported a set of whole-ecosystem nutrient fertilizations. A second grant from the US National Science Foundation would also take advantage of these experiments to test how P affects microbes with different life history strategies that might be P-intensive (the "growth rate hypothesis," Elser et al. 2000). Basically, we wanted to analyze what happens if we add P and N to CCB.

This was a simple idea that proved more difficult than expected (remember the CCB motto?). The plan was to work on an aquatic system named Pozas Rojas, a site where nine small but permanent hypersaline ponds formed a crescent moon-like circle around a larger deeper lagoon (Fig. 1.2d). Data from the site had shown not only was the site highly stoichiometrically imbalanced (N:P 159:1) but that bacteria from Pozas Rojas were fighting and "stealing" P from other organisms. This result was different from that observed for the pseudomonad-dominated microbial mats previously studied, whereas those genes related to P acquisition were underrepresented (Peimbert et al. 2012). What would happen if we gave bacteria from Pozas Rojas free P. That was the plan we had in mind. However, on April 2010 Hurricane Alex arrived in the basin, converting the isolated ponds of Pozas Rojas into a single, large, strange lagoon. Jim and his collaborators, along with the UNAM-CINVESTAV team, had to redraw their plans. We decided to change their field site to the Churince (Valeria's favorite) that by now was private property, having been protected by Carlos Slim's donation. This protection allowed them to move in and deploy equipment and maintain control on the experiment. It also helped to fulfill the ambitious "total inventory" WWF-Fundacion Carlos Slim project. In May 2011, the teams started a first experiment involving mesocosms deployed in Lagunita, a small pond near Laguna Intermedia. Because the pond was lined with deep mud that prevented sampling access, it was necessary to build and move two large metal scaffolds (8 × 4 m, welded in situ; remember again, our motto of maximum effort). It took a large team of students, friends, and locals to lift that heavy scaffold up, paint it (a bright green), and install it so that students and technicians "could walk" over water (Fig. 1.1m) to set up acrylic tube mesocosms for experimentation. There were four treatments with four replicates (P only added; both N and P added at a ratio of N:P 16:1, both N and P added at N:P 75:1; a control to which no P was added). Various limnological analyses documented how the ecosystem responded (Lee et al. 2015), while community structure was analyzed using both 16S and 18S rDNA (Lee et al. 2017). The results showed that under P fertilization, microbial communities in the water column changed abruptly due to the response of the previously rare, P-loving algae as well as fast-growing "*r*-strategist" microbes, outcompeting the slow-growing P optimizers that previously dominated the system. As expected, the quality of the extracted DNA was an issue (NASA and JCVI ended up using UNAM's extraction strategy). Hence, Zarraz Lee, ASU postdoc in charge of the experiment, had to move to Mexico to work on the DNA. A student, Gabriel Yaxal Ponce-Soto (Fig. 1.1m), got his PhD analyzing the changes of the *Pseudomona's* "fight club" or cooperation strategies before and after the nutrient experiment (Ponce-Soto et al. 2015). This was the continuation of the observations on aquatic *Pseudomonas'* "fight club" in Churince (Aguirre-von-Wobeser et al. 2013, 2015), where a similar "xenophobia" to stranger strains was demonstrated in bacteria that lived in the non-structured water column. Probably this intense xenophobia along with stochastic events can explain the very large differences in diversity from one site to the next (Espinosa-Asuar et al. 2015).

In May 2012, another fertilization experiment was conducted, but in this iteration, the acrylic tubes were used as controls, and the whole small lagoon outside the mesocosms was enriched (Okie et al. unpublished). Furthermore, replicated metagenome sequences were obtained for both water and sediment to test our "growth rate hypothesis" that proposed that taxa that are favored at high nutrient supply would have (among other things) more ribosomal copies to grow faster than the native slow growers. Analyses of the metagenomes (empowered by statistical analyses due to the replicate samples) yielded results consistent with an emerging view of a "genomic signature of growth" (Vieira-Silva and Rocha 2010): compared to the unfertilized controls, in the fertilized pond microbes proliferated that carried more copies of ribosomal RNA genes and of transfer RNA genes (to enhance translational capacity), exhibited stronger codon usage (to enhance translational efficiency), had larger genomes (since the nutrient pressures of genome construction were relieved), and higher GC content (perhaps because high-GC DNA is more costly to construct). Nutrients appear to deeply shape the structure of microbial communities at CCB and drive their evolution. Meanwhile, Eric Moody, Jim Elser's student, analyzed the stoichiometry of fishes in different systems.

As part of the test of the stoichiometric growth rate hypothesis, Jorge Valdivia (Fig. 1.1p), a PhD student at UNAM, in collaboration with Gabriela Olmedo from CINVESTAV and Jim Elser from ASU, performed a survey of the number of ribosomal operons in a collection of *Bacillus* while also assessing their growth rates and cellular C:N:P stoichiometry (Valdivia-Anistro et al. 2016). As with many things at CCB, the *Bacillus* from this site did not obey theory quite well since it seems that the number of operons was not related to the growth rate or to cell P-content in this clade. Instead, each lineage showed unique behavior, suggesting that, for these taxa, splitting the niche using different strategies was more important than optimizing of growth rate.

At the same time, Valerie de Anda (Fig. 1.1r), a new PhD student, became interested in using metagenomics to follow the decay of the ecosystem as the Churince systems, as its ponds became desiccated. Specifically, she wanted to explore the sulfur cycle that depended on the deep aquifer with magmatic influence that changed as the water level went down. Valerie used the data obtained on 2011 and also used the sampling scaffold to sample three microbial mats in Lagunita across 2 years, starting in Fall 2012. She first had to assemble the molecular pathways involved in the sulfur cycle, which was still a scientific open question (De Anda et al. 2017), and then she had to assess the community changes by analyzing 12 metagenomes, from the two seasons across the 2 years (Winter and Fall 2012–2013) (article ready for submission).

By 2012, a complete number of the journal *Astrobiology* was dedicated to CCB, including metagenomic studies done by Mexican students (the first metagenomes conducted in México) with very good quality DNA and a good coverage (Bonilla-Rosso et al. 2012; Peimbert et al. 2012). Also published in this issue was a paper on a molecular clock that used different *Bacillus* genomes from the site, also based on sequences done in Mexico (Moreno-Letelier et al. 2012). We also published a review on the biology and studies of this "Precambrian park" (Souza et al. 2012b) and a soil succession study (López-Lozano et al. 2012), along with some thoughts about space stowaways and planetary security (Siefert et al. 2012) and how microbialites were assembled (Nitti et al. 2012). Minckley's "lost ocean world" was no more just a poetic idea but a scientifically established pattern for many different lineages. It would seem as if a "niche bubble" persisted in CCB, retaining complete microbial communities as they coevolved together for a very long time, sharing the task of survival by eating any newcomer that was not part of the ancient "negotiations" (Souza et al. 2018).

Meanwhile, the total inventory of Churince project came to encompass the work of more than 100 scientists, 50 CBTA22 students, and 10 high school teachers in 6 years, producing close to 50 publications including many dissertations (Fig. 1.1t–w). As a consequence of all this work, Churince is one of the sites more intensely studied from a natural history perspective, emphasizing that most other well-studied sites do not intensively document microbial diversity. The enormous database will be soon open to the public through CONABIO and UNAM portals. The high school students are knowledgeable in molecular biology and ecology since they were part of the inventory project. Their school moved from ranking as one of the lowest (200) to among the highest rankings for rural schools, in just 4 years. Now, not only are students no longer dropping out of middle and high school; they are getting into college too. Their futures have been transformed, one of them, Hector Arocha got his PhD in Biotechnology in August 2018, his project won an award as the best project of the Universidad Autónoma de Nuevo León and he is returning to work with the kids of CBTA in bioprospecting the pozas genetics resources following the Nagoya treaty.

The education and conservation aspects are perhaps the most important result of this microbial saga. Even as Churince remains extremely endangered, these young people may save it as their parents and grandparents are beginning to change the way they think about the basin's water. The local teenagers (middle and high school) are teaching their elders twenty-first-century agriculture. Right now, they are carrying out bio-prospecting projects to uncover the biotechnological potential of CCB's microbes following to the Nagoya treaty to assure that financial benefits come to the whole community. They are taking the future in their hands using their molecular lab as the tool.

These, and other studies involving the environment, bacteria, fungi, viruses, plants, and animals will be discussed in this volume and accompanying volumes. Please enjoy your trip into the depths of Cuatro Cienegas' many mysteries.

Acknowledgments This chapter was written during a sabbatical leave of VS in the Department of Ecology, Evolution and Behavior University of Minnesota in Dr. Michael Travisano's laboratory and of LEE in the Department of Plant and Microbial Biology, University of Minnesota in Dr. Peter Tiffin's laboratory, both supported by the program PASPA-DGAPA, UNAM.

References

Aguirre-von-Wobeser E et al (2013) Two-role model of an interaction network of free-living gamma-proteobacteria from an oligotrophic environment. Microb Ecol 16(5):1366–1377

Aguirre-von-Wobeser E et al (2015) Theoretical analysis of the cost of antagonistic activity for aquatic bacteria in oligotrophic environment. Front Microbiol 6:490. https://doi.org/10.3389/fmicb.2015.00490

Alcaraz LD et al (2008) The genome of Bacillus coahuilensis reveals adaptations essential for survival in the relic of an ancient marine environment. PNAS 105:5803–5808

Alcaraz LD et al (2010) Understanding the evolutionary relationships and major traits of bacillus through comparative genomics. BMC Genomics 11:332

Avitia M et al (2015) Common population expansions in coexisting bacterial lineages revealed by genetic evidence. Peer J 2:e696

Bonilla-Rosso G et al (2012) Comparative metagenomics of two microbial mats at Cuatro Ciénegas Basin II: community structure and composition in oligotrophic environments. Astrobiology 12(7):659–673

Bonilla-Rosso G et al (2013) Understanding microbial community diversity metrics derived from metagenomes: performance evaluation using simulated data sets. FEMS Microbiol Ecol 82:37–49

Breitbart M et al (2009) Microbial communities associated with carbonate biomineralization in modern freshwater Microbialites. Microb Ecol 11:16–34

Carson EW, Dowling TE (2006) Influence of hydrogeographic history and hybridization on the distribution of genetic variation in the pupfishes *Cyprinodon atrorus* and *C. bifasciatus*. Mol Ecol 15:667–679

Carson EW et al (2012) Relationships between spatio-temporal environmental and genetic variation reveal an important influence of exogenous selection in a pupfish hybrid zone. Mol Ecol 21:1209–1222

Carson EW et al (2013) Low mitochondrial DNA sequence variation in the micro endemic Cuatro Ciénegas platyfish *Xiphophorus gordoni*. Western North Am Nat 73:224–229

Carson EW et al (2015) Mitochondrial DNA diversity and phylogeography of *Lucania interioris* inform biodiversity conservation in the Cuatro Ciénegas Casin, México. West North Am Nat 75:200–208

Cerritos R et al (2008) *Bacillus coahuilensis* sp. nov. a new moderately halophilic species from different ponds in the Cuatro Ciénegas Valley in Coahuila, México. (2008). Int J Syst Evol Microbiol 58:919–923

Cerritos R et al (2010) Diversity of culturable thermo-resistant aquatic bacteria along an environmental gradient in Cuatro Cienegas, Coahuila, México. Anton Leeuw Int J Gen Mol Microbiol 99(2):303–318. https://doi.org/10.1007/s10482-010-9490-9

Corman JR et al (2016) Interaction between lithification and resource availaibility in the microbialites of Rio mesquites, Cuatro Ciénegas, México. Geobiology 17:176–189

De Anda V et al. (2017) MEBS, a software platform to evaluate large (meta) genomic collections according to their metabolic machinery: unraveling the sulfur cycle 2017. Gigasciences 6(11):1–7 GIGA-D-17-00134

Desnues CH et al (2008) Endemism of viruses in modern stromatolites and thrombolites. Nature 452:340–342

Elser JJ et al (2000) Nutritional constraints in terrestrial and freshwater food webs. Nature 408:578–580
Elser JJ et al (2005a) Effects of phosphorus enrichment and grazing snails on modern stromatolitic microbial communities. Freshw Biol 50:1808–1825
Elser JJ et al (2005b) Response of grazing snails to phosphorus enrichment of modern stromatolitic microbial communities. Freshw Biol 50(11):1826–1835
Elser JJ et al (2006) Early Cambrian food webs on a trophic knife-edge? A hypothesis and preliminary data from a modern stromatolite-based 501 ecosystem. Ecol Lett 9(3):295–303
Escalante AE et al (2008) Diversity of aquatic prokaryotic communities in the Cuatro Cienegas basin. FEMS Microbiol Ecol 65:50–60
Escalante AE et al (2009) *Pseudomonas cuatrocienegasensis* sp. nov., isolated from an evaporating lagoon in the Cuatro Cienegas valley in Coahuila, Mexico. Int J Syst Evol Microbiol 59:1416–1420
Espinosa-Asuar L et al (2015) Aquatic bacterial assemblage structure in Pozas Azules, Cuatro Ciénegas Basin in México: deterministic vs. stochastic processes. Int Microbiol 18(2):105–115. https://doi.org/10.2436/20.1501.01.240
Gómez-Lunar Z et al (2016) Microevolution analysis of Bacillus coahuilensis unveils differences in phosphorus acquisition strategies and their regulation. Front Microbiol 7:58. https://doi.org/10.3389/fmicro.2016.0005
Hernández-Becerra N et al (2016) Agricultural land-use change in a Mexican oligotrophic desert depletes ecosystem stability. Peer J 4:e2365. https://doi.org/10.7717/peerj.2365
Lee Z et al (2015) Stoichiometrically inbalanced shallow ponds in the Cuatro Cienegas Basin is highly sensitive to nutrient enrichment. Plos One. https://doi.org/10.1371/journal.pone.0123949
Lee Z et al (2017) Nutrient stoichiometry shapes microbial community structure in an evaporitic shallow pond. Front Microbiol Ecol 8:949. ISSN: 1664-302X. FI 4.165. https://doi.org/10.3389/fmicb.2017.00949
López-Lozano NE et al (2012) Bacterial communities and the nitrogen cycle in the gypsum soils of Cuatro Ciénegas Basin, Coahuila: a Mars analogue. Astrobiology 12(7):699–709
López-Lozano NE et al (2013) Microbial secondary sucesión in soil microcosmos of a desert oasis in the Cuatro Cienegas Basin. Mexico Peer J. https://doi.org/10.7717/peerj.47
Minckley W (1969) Environments of the bolson of Cuatro Ciénegas, Coahuila, Mexico, with special reference to the aquatic biota. Texas Western Press. Univ Texas, El paso Sci Ser 2:1–65
Minckley WL (1992) Three decades near Cuatro Cienegas. Mexico: photographic documentarion and a plea for area conservation. J Ariz-Nev Acad Sci 26:89–118
Moreno-Letelier A et al (2011) Parallel evolution and horizontal gene transfer of the *pst* operon in *Bacillus* from oligotrophic environments. Int Evol Biol 2011:1–10. https://doi.org/10.4061/2011/781642
Moreno-Letelier A et al (2012) Divergence and phylogeny of *Firmicutes* from the Cuatro Cienegas Basin, Mexico. Astrobiology 12(7):674–684
Nitti A et al (2012) Spatially resolved genomic, stable isotopic, and lipid analyses of a modern freshwater microbialite from Cuatro Ciénegas, Mexico. Astrobiology 12(7):685–698. https://doi.org/10.1089/ast.2011.0812
Pajares S et al (2012) Mesocosms of aquatic bacterial communities from the Cuatro Cienegas Basin (Mexico): a tool to test bacterial community response to environmental stress. Microbiol Ecol 64:346–358
Pajares S et al (2013) Response to long-term environmental stress of aquatic bacterial communities from the Cuatro Cienegas Basin (Mexico). Antonie Van Leeuwenhoek 104:1159–1167
Pajares S et al (2015) Multivariate and phylogenetic analyses to assess the response of bacterial mat communities from an ancient aquatic ecosystem to different scenarios of climate change. PLoS One 10(3):e0119741. https://doi.org/10.1371/journal.pone.0119741
Peimbert M et al (2012) Extreme Redfield ratios, metagenomic and microbial diversity analyses of a seasonal shallow red pool in Cuatro Cienegas Coahuila, Mexico. Astrobiology 12(7):648–658

Pérez-Gutiérrez RA et al (2012) Antagonism influences assembly of a Bacillus guild in a local community and is depicted as a food-chain network. ISME J 7:487. https://doi.org/10.1038/ismej.2012.119

Perroni Y et al (2014) Relationship between soil P fractions and microbial biomas in an oligotrophic grassland-desert scrub system. Ecol Res 29:463–472

Perroni Y et al (2015) Plant species identity and soil P forms in an oligotrophic grassland-desert scrub system. J Arid Environ 108:29–37

Ponce-Soto GY et al (2015) Enrichment experiment reduces diversity and changes microbial interactions in an ultra-oligotrophic environment. Front Microbiol 6:246

Rebollar EA et al (2012) Water–sediment niche differentiation in ancient marine lineages of Exiguobacterium endemic to the Cuatro Cienegas Basin. Environ Microbiol 14:2323–2333

Rodríguez-Verdugo A et al (2012) Diversity across seasons of Culturable pseudomonas from a desiccation lagoon in Cuatro Cienegas, Mexico. Int J Microbiol 2012:1–10. https://doi.org/10.1155/2012/201389

Siefert JL et al (2012) Microbial stowaways: inimitable survivors or hopeless pioneers? Astrobiology 12(7):710–715

Souza V et al (2006) An endangered oasis of aquatic microbial biodiversity in the Chihuahua desert. PNAS 103:6566–6570

Souza V et al (2008) Microbial endemism: does extreme nutrient limitation enhance speciation? Nat Rev Microbiol 6:559–564

Souza V et al (2012a) The Cuatro Cienegas Basin in Coahuila, Mexico: an astrobiological Precambrian park. Astrobiology 12(7):641–647

Souza V et al (2012b) Travel, sex, and food: what's speciation got to do with it? Astrobiology 12(7):634–640

Souza V et al (2018) The lost world of Cuatro Cienegas Basin, a relictual bacterial niche in a desert oasis. Preprint en Biorxvi, acepted with corrections in eLife

Tapia-Torres Y et al (2015) Ecoenzymatic stoichiometry at extremes: how microbes cope with an ultraoligotrophic desert soil. Soil Biol Biochem 87(2015):34–42

Tapia-Torres Y et al (2016) How to live with phosphorus scarcity in soil and sediment: lessons from bacteria. Appl Environ Microbiol 82(15):4652–4662. https://doi.org/10.1128/AEM.00160-16

Toribio J et al (2011) Characterization of a novel biosurfactant producing Pseudomonas koreensis lineage that is endemic to Cuatro Ciénegas Basin. Syst Appl Microbiol 34:531–535

Valdivia-Anistro J et al (2016) Variability of rRNA operon copy number and growth rate dynamics of bacillus isolated from an extremely oligotrophic aquatic ecosystem. Front Microbiol 6:1486. https://doi.org/10.3389/fmicb.2015.01486

Vieira-Silva S, Rocha EPC (2010) The systemic imprint of growth and its uses in ecological (meta) genomics. N.A. Moran [ed.]. PLoS Genet 6:e1000808

Chapter 2
In the Beginning, There Was Fire: Cuatro Ciénegas Basin (CCB) and the Long History of Life on Earth

Valeria Souza and Luis E. Eguiarte

Abstract To understand the biodiversity of the Cuatro Cienegas Basin (CCB), we realized that some historical context was needed. In this case, the history reaches far back in geological time. This chapter will be a "once upon a time" story—told using the available data in the geological record—that follows the ever-increasing complexity of metabolic life and how this complexity altered niches have shaped the Earth's environment. To really kick off the metabolism saga, we will start when the planet was so young it did not yet have a moon, and we will imagine several origin-of-life scenarios. We will explore everything from the origin of methanogens in the early Archean eon to the buildup of amazingly biogeochemically complex microbial mats that singlehandedly constructed a blue planet by releasing O_2 as a by-product of the quantic reaction of photosynthesis. We will also explain how the Cryogenian period at the end of the Precambrian changed the nutritional equation, opening an unchartered niche of high phosphate (PO_4) levels to the early eukaryotes in the Ediacaran, thus allowing the evolution of animals, plants, and fungi. We also suggest that the microbial communities present along the shallow oligotrophic coasts are preserved, in an ancient niche of a world without available P. These ancient niches still persist today in at least one site in the world—within the Chihuahuan Desert in what is now Coahuila, a hyper-diverse oasis that for millions of years was a shallow sea—the Cuatro Ciénegas Basin.

Introduction

As we already explored in the first chapter, the Cuatro Ciénegas Basin (CCB), in the North of Mexico, is an oasis containing a unique and amazing biological diversity dominated by bacteria. These bacteria are not only very diverse in the aquatic

V. Souza (✉) · L. E. Eguiarte
Laboratorio de Evolución Molecular y Experimental, Departamento de Ecología Evolutiva, Instituto de Ecología, Universidad Nacional Autónoma de México, México City, Mexico
e-mail: souza@unam.mx

V. Souza et al. (eds.), *Cuatro Ciénegas Ecology, Natural History and Microbiology*, Cuatro Ciénegas Basin: An Endangered Hyperdiverse Oasis,
https://doi.org/10.1007/978-3-319-93423-5_2

environments of the basin (in the water of every pond, river, and lagoon and in the sediments, microbial mats, and stromatolites therein) but also in the soils. These environments are rich with unique bacteria and display high levels of community turnover; the bacterial communities in one site are completely different to those that exist just a few meters away. Not only that, we now have strong evidence that the bacteria found in CCB usually are very different from bacteria found anywhere else in the world. Even after nearly 20 years of intense environmental microbiological studies across the world, the genetic sequences of bacteria from CCB remain divergent from any other known relative. It has been repeatedly demonstrated that many bacterial clades in CCB represent very old lineages that have been separated from their relatives for up to hundreds of millions of years. As Carl Sagan would say: "extraordinary claims require extraordinary evidence." Thus, along this book, we will describe the diversity of bacteria in CCB and explore the extraordinary evidence we have accumulated that supports our claims of bacterial uniqueness in CCB. In the other books of this series, we will explore other aspects of CCB's biological uniqueness and diversity.

But first, in this chapter we will review the history of Earth, from its very early beginnings to the Anthropocene: the era of men. We will connect this long geological history with the hyper-diverse bacteria of CCB using modern ecological niche theory.

Earth's Early History and the Origin of Life

In the beginning, there was fire: magmatic fire caused by the collision of two planetoids early in the solar system's origin 4.5 billion years ago. Terra and Theia were two young planets, only 100 million years old, that fell into the same orbit and onto a collision course with each other. In the end, we were lucky. This cosmic collision gave the Earth its moon, a larger planetary mass that could hold an atmosphere, and a second opportunity for an ocean according to the latest theory about the chimeric nature of our planet and the moon (Young et al. 2016). However, nothing could have lived through such a high impact, and so our story begins later, but surprisingly, not too much later.

Water is a key ingredient for life and is the most abundant molecule in the universe. It is not only part of the stardust that originally formed the planets but is also particularly abundant in the stars' magmatic material. Water is what gives comets their tails and meteorites their ice (Russel et al. 2017). Asteroids in particular—cosmic bodies formed from the debris of inner planetary formations—contain the most water of all types of meteorites, particularly the asteroids known as carbonaceous chondrites (Russel et al. 2017). These asteroids not only carry water but, according to isotopic evidence, also contain a large load of organic material synthesized during their cosmic journey, fed by energy from the stars (Alexander et al. 2012).

So, what of the water and organic material on our Earth? Between 4 and 3 billion years ago, asteroids constantly fed the inner planets of our solar system with water and organic molecules through collisions. This era of heavy bombardment created a rich organic input of abiotic material to Earth. This was also a time when most of the moon's and inner planets' craters were formed. However, this hypothesis of abundant meteorite cataclysms caused by rogue pieces of interstellar material was conflicting (Abranov and Mojzsis 2009), until recent evidence of early life was found.

The panoply of prebiotic organic material that was dissolved in the water—the so-called prebiotic soup—was synthesized both locally on Earth and celestially on the asteroids. This prebiotic soup was most likely very rich in carbon (C), nitrogen (N), sulfur (S), and iron (Fe), along with other inorganic compounds, though the latter is still controversial.

Phosphorus (P), a key element for life's energy and for storing information, was probably limited in the early ocean sediments due to its reactivity with calcium (Ca) and Fe (Kipp and Stüeken 2017). Interestingly, despite its scarcity since the beginning, this element was a fundamental backbone of prebiotic life. Prebiotic life is usually suspected to have existed in an RNA world (Lazcano and Miller 1996). Even then, P was an essential element used for constructing nucleic acids (RNA) and ATP: the universal, and probably prebiotic, form of chemical energy. Phosphorus is indeed special, given its particular energy content in the bond between two PO_4. It is also possible that P was not so limited in the ancient rocks. It was rare, but abundant enough for the early "experiments" of life (Kipp and Stüeken 2017). Such experiments included the construction of cells, energetic metabolism, and the creation of DNA and other protein synthesis complexes with an established genetic code.

Genomic evidence points out that LUCA (last universal common ancestor) had at least 500 genes. LUCA and its descendants most likely evolved in a very local site; thus, it would have taken quite some time for the oceans' total biomass to increase enough to deplete the early and easily available P. The niche, in the beginning, was simple and depended mostly on physicochemical constraints.

Recent findings placed the earliest signatures of life close to 4 billion years ago (bya), just at the end of the Hadean (Tashiro et al. 2017; Dodd et al. 2017)—a period when, despite the occasional collision of meteorites, the temperature of the sea was more or less constant (Valley et al. 2002). In all cases, that evidence points to early methanogens. In turn, earlier evidence of microbial mats on Earth got pushed back to 3.7 bya (Nutman et al. 2017), suggesting that the origin of life after its prebiotic chemical evolution stage happened very quickly once the oceans were cool enough to be liquid. Not only that, earlier complex communities did not need a long time to develop. Those life experiments survived. We are living proof that life transforms everything around it.

These early complex microbial communities developed under the ocean water, escaping harmful UV light. At this point in time, the protective ozone layer was not yet formed as O_2 was rare. The early atmosphere in the Archean was a product of magmatic degasification, as well as the interaction between the early oceans and flowing magma (Russel et al. 2017). Hence, the Hadean atmosphere was dominated by carbon dioxide, methane, and nitrogen (in much the same ratio as the present-day atmospheres of Venus and Mars), along with some NO, CO, P_4O_{10}, SO_2, and native sulfur from the interaction with magmatic materials. Due to the resulting methane-sulfur haze, both the ocean and the atmosphere were of an orange-yellow hue.

Early Microbial Communities: Microbial Mats and Stromatolites

The early life had an entire planet to conquer. However, since the beginning, life was constrained by the extreme heat of volcanic magma, the intensity of unfiltered UV light, and the presence of nasty neighbors. All these variables limited the early ecological niches for life under water, protected by the filtering capabilities of the sea though even these habitats could be wiped out by an exploding volcano or a falling meteorite. Early on, nutritional codependence made sense, as nutrients can be more efficiently degraded when done in tandem or by cross-feeding (Morris et al. 2012). Hence, it is possible that multiple species of microbes started to associate with each other in what is known as microbial mats (Fig. 2.1a, b). Microbial mats developed complex communities, as can be seen in the early fossil record, and look very much like their modern-day equivalent (De Anda et al. 2017). These communities built up, layer by layer, creating the complex metabolic food web that transformed the planet. Depending on the presence or absence of calcium carbonate in their local site, some of these mats became thicker and started to precipitate calcium as part of their metabolic activities, a behavior usually associated with anaerobic photosynthesis. These microbial mats are named after their physical appearance, known in the geological literature as stromatolites (i.e., banded rocks). Living microbial mats and stromatolites (Fig. 2.1) still can be found on Earth, usually developing only in extreme environments where algae do not grow. Algae are fast-growing competitors and are the definition of "nasty neighbors," as they dominate the Sun's light over the slow-growing stromatolites.

Let's analyze a microbial mat carefully (Fig. 2.1a). Layer by colorful layer provides an explanation of the evolution of metabolisms on early Earth and what we call—poetically—the assemblage of the "clock of life."

Layer 1: the black methanogens represent the first metabolism with a fossil record, as can be seen in the isotopic signature of enriched C^{14} in the zircon diamonds of 4 billion years ago (Dodd et al. 2017; Tashiro et al. 2017). This is a simple metabolism that used CO_2 from the atmosphere and H from minerals in order to produce methane and obtain ATP. This form of metabolism occurred via a process called chemiosmosis and only occurred under strictly non-aerobic conditions. Carbon is the terminal electron acceptor in modern methanogens and comes from acetic acid or CO_2. The purpose of the reaction is to create a membrane potential that allows ATP synthase to rotate and attach a phosphate to ADP, producing the energy-rich ATP molecule. This energy is used to combine more C

Fig. 2.1 Microbial mats and stromatolites at the Cuatro Cienegas Basin. (**a**) Microbial mat from Churince, (**b**) microbial mat from the "domes of the archean," (**c**) stromatolite reef in Pozas Azules, PRONATURA ranch, and (**d**) microbial mat in pozas rojas

and construct small hydrocarbons and methane as well as all cell precursors. Nowadays, the dominant lineages capable of fixing CO_2 into methane are the Archaea, a distinct domain of cells without nuclei (prokaryotes). However, the users of this metabolism, the methanotrophs that also live in that black first layer are Bacteria, the other deep lineage of prokaryotes. Then, the ancestral black layer suggests that early on life had already diverged into two very different domains—domains so different that they have different ribosomes (that is how Woese and Fox discovered them in 1977) and even different membrane structures. To illustrate this difference, we can look to the specific lipid signature of Archaea found in 3.8 billion years old rocks from Greenland (Hahn and Haug 1986). This signature is the L-glycerol-ester bond in Archaean phospholipids that is specific to their membrane structure and cell wall composition. Bacterial membrane structures, on the other hand, involve a D-glycerol-ether bond, and their cell wall composition requires the polymer peptidoglycan, which archaeans lack. There are some Archaea, the Methanobacteriales, which can either use formate to reduce CO_2 or use methanol to produce methane. Of particular interest is that these Archaea have pseudopeptidoglycan cell walls, which confirms an early splitting of these two older domains of life.

Layer 2: this beige layer is rich in organisms that gain their energy predominately from metabolizing sulfur. This type of metabolism is used by many different lineages of heterotrophic and lithotrophic bacteria that generate energy from "rocks and minerals" (De Anda et al. 2017). In this diverse layer, we have four groups: (1) sulfate-reducing and elemental sulfur-reducing bacteria, (2) elemental sulfur-reducing and sulfur-oxidizing bacteria, (3) sulfate-reducing bacteria, and (4) the colorless sulfur bacteria. All of them are present within this layer in a gradient of redox possibilities given by the mat structure in anoxic conditions. Other bacteria that live in this layer are the nitrifying bacteria, iron oxidizers, and hydrogen oxidizers. These ancient autotrophs are inefficient and slow, but back then life was in no hurry. In fact, methanogens are even slower!

Layer 3: this amazing layer displays vibrant colors: intermingled deep greens and purples. These colors are the essence of the early "invention" of photosynthesis by the codependent purple and green bacteria. Both lineages use sulfur and non-sulfur organic compounds as electron acceptors. The early photoreceptors that used low-energy light to obtain energy were derived from simple carotenes. Carotenes are pigments that evolved very early as protectors against the Sun's UV light. From these carotenoids, light antennas evolved (generally called "bacteriochlorophylls"). The difference between the purple and deep green photosystems is that the sulfur green bacteria have type I reaction centers with a wavelength receptor P840, while the purple uses the type II reaction center with a wavelength receptor of P870; this allows them to "pick" different parts of the light spectrum and coexist. Moreover, to understand how these photosystems are constructed is relevant, since the antennae are associated in all cases to the ATP synthase.

But first, it is important to understand how photosynthesis generates energy. The Sun produces photons that, when absorbed by plants, excite electrons within the plant's pigments. Photosynthesis is all about harnessing the energy from the excited electrons to produce ATP. So how did the earliest photosynthetic cells harness an excited electron? They needed a system that extended from the antenna but also penetrated the membrane to the LH (light-harvesting) proteins. These LH proteins feed the excited electrons to the reaction center of the pigment, where the electron is "shepherded" by cytochrome proteins. These electron transfer proteins then trigger a photochemical reaction that separates positively charged protons and negatively charged atoms. This separation is important because the protons are the "fuel" that causes the ATP synthase to rotate and attach a phosphate to ADP, creating an energy-rich ATP molecule. This energy is used to bond C atoms that enable the construction of glucose. The light used by the microbes within this layer is close to infrared, suggesting that they evolved in not-so-superficial layers of the ocean. Most likely, these microbes lived in a deeper photic zone that protected them from dangerous UV light, since there was no ozone to protect life on a mostly orange-colored anoxic planet (Russel et al. 2017).

Layer 4: the one that changed the world forever—is the blue-green layer. Bacteria in this penultimate layer transformed the atmosphere bubble by bubble, increasing the amount of atmospheric oxygen and later precipitating a global glaciation that changed the anoxic nature of the oceans' sulfur-rich chemistry. The extraordinary complexity of oxygenic photosynthesis needs a bit more "storytelling" since there is no fossil evidence to fill the gaps.

First, we have fossil evidence of local oxygen in the deep ocean during the Archean eon (Lyons et al. 2014), suggesting the presence of this new niche as well as the evolution of high-energy photosynthesis in the *Cyanobacteria* lineage. Before that, there was possibly some oxygen being stored in the bottom of the sea within the clay and pyrite found in hydrothermal vents, where, locally, heat could break apart water molecules. We believe that it was in these vents where oxygen tolerance must have evolved, which then opened up the possibility of blue-green photosynthesis. Let's remember that life evolved mostly without oxygen and that this element is very electronegative and a powerful oxidizer of DNA, despite being the most efficient electron acceptor in the respiration of efficient aerobic bacteria.

Cyanobacteria have not one but two types of photosystems acting in synchrony in a Z type of "electron dance." These systems together generate a large amount of ATP allowing the cyanobacteria to capture more carbon, which they then store in special structures called carboxysomes. These sites are where the Calvin cycle (the one that constructs glucose out of 6 CO_2 molecules and water) occurs. Cyanobacteria also have their membrane folded, forming thylakoids. This increases the membrane surface area, which allows a large number of photosystems (I and II) to be housed, each one containing the high-energy light antennas chlorophyll a and b.

Since the Earth crust cooled 2.5 bya (Fig. 2.2), the gas composition of the atmosphere changed as greenhouse gases diminished, thanks to decreased volcanic activity. Unfortunately, methanogens and sulfur bacteria that depended on magmatic metals suffered. The Archean ocean crust was thicker and hotter (down to 20 km) than the present crust, making it harder and generally more resistant to subduction.

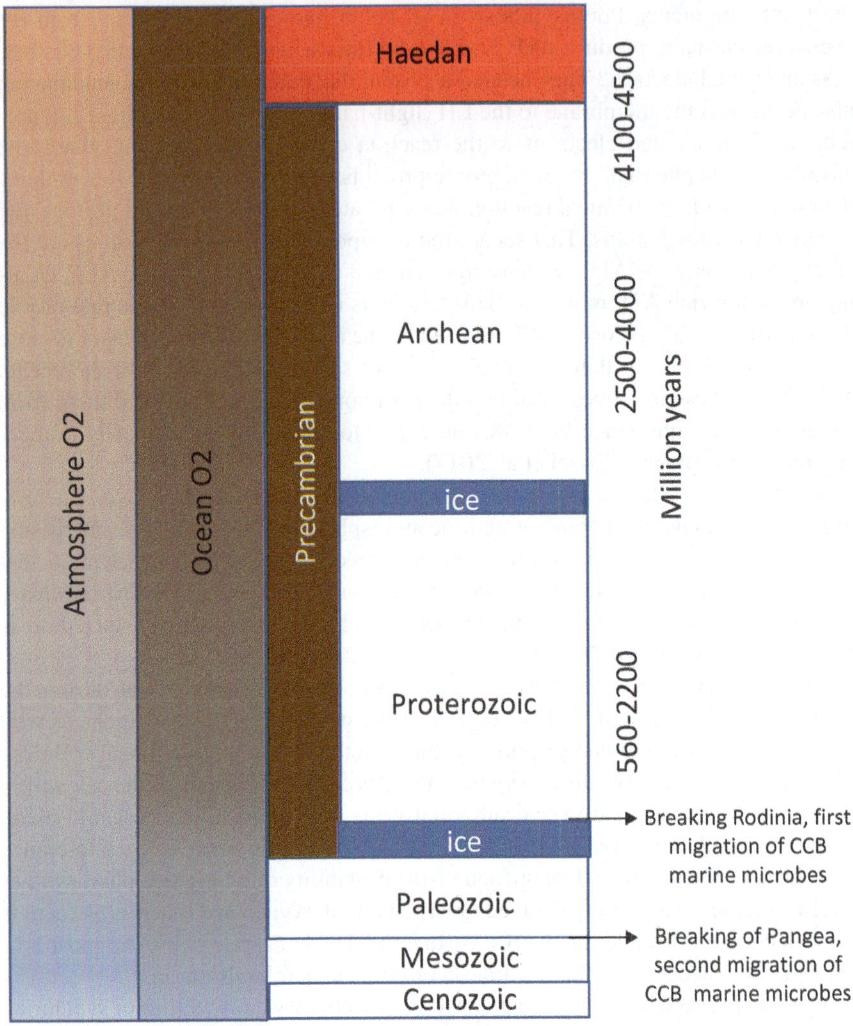

Fig. 2.2 Both column show an estimate of oxygen, the first one in the atmosphere and the second in the ocean as suggested in shades of blue. Note that while the atmosphere oxygen slowly increased after the late Archean glaciation despited as a blue line, the ocean oxygen is correlated with the late Proterozoic glaciation also despited as a blue line. Arrows show the two breakage of super continents that gave origin to marine microbes of CCB

However, some subduction occurred, and a few of these subduction zones generated what is called the Archean greenstone belt (Sleep and Windley 1982). By the end of the Archean eon, the core granite cratons of the continents had already formed, and volcanic activity diminished as the crust began to cool and thin.

According to the evolution of stars similar to the Sun, the Sun's light was 70% fainter 4 bya. While the crust cooled down, however, the greenhouse gases we just described kept the young ocean of Earth warm until the Huronian ice age—the first global glaciation that marked the end of the Archean eon (Tang and Chen 2013).

The Huronian ice age, caused by reduced volcanic activity and a fainter Sun, created a "Snowball Earth" that took hold of most of the planet and marked the boundary between both eras within the Proterozoic (Kirschvink et al. 2000). All the available rocks from that event, spanning a 400-million-year period, are scratched with glacial erosion. Microbial mats and stromatolites did not survive under the ice of Snowball Earth since they need liquid water and energy to function. However, they survived a small non-frozen band of the ocean at the equator, and it is within this band that these microbial communities survived the Ice age.

It is possible that this glaciation event was interrupted by a buildup of granite continental material that created the first supercontinent, Nuna, and began the tectonic cycle. Nuna had shores and shallow seas that opened new areas and ecological niches for stromatolites and microbial mats (Spencer et al. 2018). Luckily for life, this heated the oceans and melted the ice. At the same time, the ice eroded the continental shield causing a release of a substantial amount of phosphate (Papineau 2010). The post-glaciation rocks exhibit banded iron formations providing evidence of the onset of the Precambrian period and suggesting the expansion of the available niches for stromatolites and microbial mats (once the oceans recovered their warmer pre-glaciation temperatures). The golden age of stromatolites had started—and it lasted more than a billion years.

Another consequence of the crust cooling down was that bioavailable nitrogen was harder to get. Genetic evidence shows that nitrogenase (*nif*) genes (the genes that allow organisms to "fix" nitrogen; Raymond et al. 2004) evolved probably in a single lineage of Bacteria that lived on an anoxic Earth. At the onset of the Precambrian, however, these genes spread across multiple different lineages of both Archaea and Bacteria by horizontal gene transfer. One hypothesis of this widespread genetic transmission is that as large-scale ice cover dramatically shrunk available niches, the greater need for SOS genetic repair increased the rate of recombination between members of the surviving microbial mats and stromatolites. Metabolic innovation was a response to the extreme environmental stress that Snowball Earth imposed upon the surviving microbial communities.

Among the possible recipients of the *nif* genes are the cyanobacteria. Many lineages in this phylum are able to fix nitrogen despite oxygen inhibiting the enzyme reaction site. At this site, NH_3 is obtained by breaking three strong N_2 covalent bonds using 16 ATP, then adding H. The problem of fixing nitrogen despite oxygen inhibition is circumvented in a couple of ways: some unicellular cyanobacteria fix nitrogen only by night (i.e., when photosynthetic activity is halted), while other filamentous cyanobacteria have special cells name heterocysts, within which photosynthesis does not occur. As you can see, cyanobacteria are very sophisticated. Their abilities evolved only once in the history of life on Earth, and in fact, plants only evolved because they engulfed cyanobacteria into their own cells to create chloroplasts.

Layer 5: the final layer—the eukaryote layer—one that took longer to evolve since we don't have glimpses of it until the Precambrian eon when oxygen began to build up in the atmosphere. Eukaryotes arose due to endosymbiosis (Margulis 1996) between different species of Achaea and bacteria. This cellular symbiosis probably occurred as part of the biogeochemical perturbation caused by the defrosting of

Snowball Earth 2.2 bya, when glacial erosion of continental rocks abruptly increased the P content of the sea (Papineau 2010). We really don't know when this occurred, since we don't have good fossil records of these early nuclei cells. We do know, however, that aerobic Archaea and bacteria could live off the oxygen bubbles emitted by cyanobacteria in the fourth layer of the stromatolites. At a time when the ocean was a soup of sulfuric acid, this fourth layer was the only possible niche for the aerobes, since all eukaryotes have mitochondria borne of aerobic bacterial endosymbionts; early eukaryotes were limited to this narrow niche since they all needed oxygen to respire.

There is another sharp increase in P in the fossil record; this one marked the end of the Precambrian and the onset of the Cambrian, and as in the first P increase, it was caused by the weathering of continental rocks caused by the ice. However, this second Snowball Earth was caused not by the decrease in volcanic activity but by the increase of photosynthesis (Brocks et al. 2017). The Cryogenian is a glaciation event that occurred when, 850 million years ago, the breakage of the supercontinent Rodinia provided more shallow shores to the microbial mats and stromatolites that had already evolved their fifth layers. Laurentia, the core of North America, was not an exception, and its shores were covered by stromatolite reefs (this is the first time that the marine bacteria from CCB diverged from its marine ancestors) (Moreno-Letelier et al. 2012; Souza et al. 2018). Carbon fixation and subsequent carbon storage by these stromatolite reef communities precipitated a sharp decrease in atmospheric CO_2. The reduction of this important greenhouse gas brought about the Sturtian glaciation, followed by a brief interglacial period characterized by high environmental P content. This opened a new niche for the P-thirsty eukaryotes, particularly for algae. The algae, embracing this eutrophication, formed a massive algal bloom that precipitated the Marinoan glaciation. Yet again, more P was introduced to the environment through glacial erosion, and with that the Ediacaran era began, 635 million years ago, marking the origin of animals. This paved the way for the Cambrian explosion, which occurred 100 million years later, once the algae made the planet blue by oxygenating the ocean (Fig. 2.2, notice the gradient of blue takes longer in the ocean than in the atmosphere).

By allowing a safe niche for the first eukaryotes, stromatolites had created their own doom. Algae grow extremely fast with abundant P, so while early animals grazed on the stromatolites, algae blocked them from drawing energy from the Sun (Elser et al. 2006). This predation/competition combo (accelerated by high nutrient content) shrunk the stromatolite's oligotrophic niche, driving them to near extinction by the Permian era. By then, the western shore of Laurentia, the site where another supercontinent—Pangea—broke 200 million years ago (Kershaw 2017), was the only place where some stromatolite fossils remained. Probably, because this particular site was poor in P, the condition that stromatolites prefer as algae cannot outcompete them in this oligotrophic niche. CCB became the western shores of the Tethys sea (Souza et al. 2006).

More Fire and the Birth of an Ocean

The tectonic breaking of Pangea was caused, like some of its predecessors, by magmatic fire and occurred mostly during the Jurassic period, when the dominant mineral in the oceans was gypsum (calcium sulfate dihydrate) instead of modern-day calcium carbonate. Hence, the geological mark of the Jurassic is gypsum, the mineral among the geological havoc that enabled the making of shells in this period.

As Pangea broke, the Jurassic Panthalassa Ocean entered CCB and never left. In fact, this is still evident today, as the western part of the basin is characterized by a gypsum-dominated ecosystem. It is also the geologic locality that corresponds to where the CCB area was formed at the onset of the Mesozoic (Souza et al. 2006; Wolaver et al. 2013). We believe that the original oligotrophic conditions of these shores remained, hosting the ancient niche where the genomes of the ancestral communities—the parallel "script" of life on Earth, the original play that early life enacted, extracting the essential elements from any possible source—are still reflected today in the genomes of the many extant bacterial communities of CCB.

The Uniqueness of Cuatro Cienegas

The large panoply of marine microbial communities survived through numerous meteor bombardments and the harsh environmental conditions of both Snowball Earths. They also survived in this particular place all the abrupt changes in environmental conditions that caused all major extinction events that wiped out most of the macroorganisms in the limits between eras. The microbial communities of CCB not only survived, but they also managed to thrive and innovate; they adapted and evolved new lineages unique to their isolated ultra-oligotrophic site. By combining different metabolisms within a single community, they produced massive amounts of oxygen, which transformed the Earth from a hazy orange planet into the blue planet full of life that we know today. These stromatolites have also survived the invasion of multicellular organisms, great extinction events, ocean level changes, and plate tectonics. They survived on the shores of the Panthalassa Ocean (now the Pacific) that became part of the proto-Atlantic in the western sea belt. They survived the uplift of the Sierra Madre Oriental, which isolated them from the ocean to a butterfly-shaped oasis in a desert basin. These unique microbial mats can either materialize as stromatolites or stay as soft non-calcareous matrices. Both strategies are successful and produce stable ecological communities that can survive and have survived millions of years of environmental disturbance (Fig. 2.1). This is due to their ability to construct their own "niche space" and to their unique capacity to be self-sustaining—performing metabolic complementation reactions with overlapping ecological functions at a microscopic scale so long as they have water, sulfur, and sunlight (De Anda et al. 2017; Des Marais 2003; Pajares et al. 2012; van Gemerden 1993).

Nowadays, stromatolites and the P-deprived environments they inhabit are like natural "time machines," as they exist only in sites that mimic past conditions. One of these "time machines" is the Cuatro Ciénegas Basin (CCB), an extremely low-P aquatic environment in a naturally isolated valley in the Chihuahuan Desert (Coahuila, México), (Souza et al. 2008, 2012).

CCB has the highest level of endemism in North America, comparable to that of the Galápagos, and is listed as a Wetland of International Importance under the International RAMSAR Convention. It is a protected natural area (ANP) for flora and fauna under the Mexican government and a priority site for Conservation of Nature, the World Wide Fund for Nature, and UNESCO.

The CCB system receives water input from deep aquifers as well as rainfall replenishment, resulting in several aquatic ecosystems each with their own unique physicochemical conditions (Wolaver et al. 2013). In recent decades, CCB has seen reduced deep-water inputs associated with agricultural water-use practices for grazing and feeding livestock. These practices initiated a drying process in one of the principal aquatic systems within CCB named "Churince" (Souza et al. 2007). This anthropogenic disturbance has led to an enormous shift in the overall ecological condition of the system as well as its constituent parts (e.g., the microbial mats). The next chapter in this book will elaborate on the present niche, the weather patterns, and the conditions of the CCB.

In the rest of this book, you will see different groups of scientists describing the conditions, adaptations, and diversity of different selected bacterial groups that have been analyzed in CCB, detailing why this particular butterfly-shaped basin is relevant to both modern science and natural conservation.

Acknowledgments This chapter was written during a sabbatical leave of VS in the Department of Ecology, Evolution, and Behavior at the University of Minnesota in Dr. Michael Travisano's laboratory and of LEE, in the Department of Plant and Microbial Biology, University of Minnesota, in Dr. Peter Tiffin's laboratory, both supported by the program PASPA-DGAPA, UNAM.

References

Abranov O, Mojzsis SJ (2009) Microbial habitability of hadean earth during the late heavy bombardment. Nature 459:419–422

Alexander CM et al (2012) The provenances of asteroids and their contributions to the volatile inventories of the terrestrial planets. Science 337:721–723

Brocks JJ et al (2017) The rise of algae in Cryogenian oceans and the emergence of animals. Nature 548:578–581

De Anda V et al (2017) MEBS, a software platform to evaluate large (meta)genomic collections according to their metabolic machinery: unravelling the sulfur cycle. GigaScience 6:1–17

Des Marais DJ (2003) Biogeochemistry of hypersaline microbial mats illustrates the dynamics of the modern microbial ecosystems and early evolution of the biosphere. Biol Bull 204:160–167

Dodd MS et al (2017) Evidence for early life in Earth's oldest hydrothermal vent precipitates. Nature 543:60–64

Elser JJ et al (2006) Trophic knife-edge: a hypothesis and preliminary data from a modern stromatolite based ecosystem. Ecol Lett 9:295–303

Hahn J, Haug P (1986) Traces of archaebacterial in ancient sediments. Syst Appl Microbiol 7:178–183

Kershaw S (2017) Paleogeographic variation in the Permian-Triassic boundary microbialites: a discussion of microbial and ocean processes after the end Permian mass extinction. Palaeogeography 6:97–107

Kirschvink JL et al (2000) Paleoproterozoic snowball earth: extreme climatic and geochemical global change and its biological consequences. PNAS 97:1400–1405

Kipp MA, Stüeken EE (2017) Biomass recycling and Earth's early phosphorus cycle. Sci Adv 3:eaao4795

Lazcano A, Miller SL (1996) The origin and early evolution of life: prebiotic chemistry, the pre-RNA world and time. Cell 85:793–798

Lyons TW et al (2014) The rise of oxygen in Earth's early ocean and atmosphere. Nature 506:307–315

Margulis L (1996) Archaeal-Eubacterial mergers in the origin of Eukarya: phylogenetic classification of life. PNAS 93:1071–1076

Moreno-Letelier A et al. (2012) Divergence and phylogeny of Firmicutes from the Cuatro Cienegas Basin, Mexico. Astrobiology 12(7):674–684

Morris JF et al (2012) The black queen hypothesis: evolution of dependencies through adaptive gene loss. MBio 3:e00036–e00012

Nutman AP et al (2017) Seeing through the magnetite: reassessing Eoarchean atmosphere composition from Isua (Greenland)>3.7 Ga banded iron formation. Geosci Front 8:1233–1240

Pajares S et al (2012) Mesocosms of aquatic bacterial communities from the Cuatro Cienegas Basin (Mexico): a tool to test bacterial community response to environmental stress. Microb Ecol 64:346–358

Papineau D (2010) Global biogeochemical changes at both ends of the Proterozoic: insights from phosphorites. Astrobiology 10:165–181

Raymond J et al (2004) The natural history of nitrogen fixation. Mol Biol Evol 21:541–554

Russel SS et al (2017) The origin, history and role of water in the evolution of the inner solar system. Phil Trans R Soc A 375:20170108

Spencer CJ et al (2018) A Paleoproterozoic tectono-magmatic lull as a potential trigger for the supercontinent cycle. Nat Geosci 11:97–101

Sleep NH, Windley BF (1982) Archean plate tectonics: constraints and inferences. J Geol 90:363–379

Souza V et al (2006) An endangered oasis of aquatic microbial biodiversity in the Chihuahua desert. PNAS 103:6566–6570

Souza V et al (2007) Protecting a window into ancient earth: towards a Precambrian park at Cuatro Cienegas, Mexico. The Citizen's Page, Evolutionary Ecology Research. http://www.evolutionary-ecology.com/citizen/citizen.html http://www.evolutionary-ecology.com/citizen/citizen.html

Souza V et al (2008) Microbial endemism: does phosphorous limitation enhance speciation? Nat Rev Microbiol 6:559–564

Souza V et al (2012) Travel, sex and food: what speciation got to do with it. Astrobiology 12:634–640

Souza V et al (2018) The lost world of Cuatro Cienegas Basin, a relictual bacterial niche in a desert oasis. Preprint en Biorxiv, accepted with corrections in eLife

Tang H, Chen Y (2013) Global glaciations and atmospheric change at ca. 2.3 Ga. Geosci Front 4:583–596

Tashiro T et al (2017) Early trace of life from 3.95 Ga sedimentary rocks in Labrador, Canada. Nature 549:516–518

Valley JW et al (2002) A cool early earth. Geology 30:351–354

van Gemerden H (1993) Microbial mats: a joint venture. Mar Geol 113:3–25

Woese CR, Fox GE (1977) Phylogenetic structure of the prokaryotic domain: the primary kingdoms. PNAS 74:5088–5090

Wolaver BD et al (2013) Identifying origins of and pathways for spring waters in a semiarid basin using He, Sr, and isotopes: Cuatro Cienegas Basin, Mexico. Geosphere 9:113–125

Young ED et al (2016) Oxygen isotopic evidence for vigorous mixing during the moon-forming giant impact. Science 351:493–496. https://doi.org/10.1126/science.aad0525

Chapter 3
The Climate of Cuatro Ciénegas Basin: Drivers and Temporal Patterns

Cristina Montiel-González, Francisco Bautista, Carmen Delgado, and Felipe García-Oliva

Abstract The Cuatro Ciénegas Basin (CCB) is considered an important biodiversity hot spot despite its arid climate conditions. The valley is located in the southern part of the Chihuahuan desert at 26° 50′ 41″ N and is strongly affected by a divergent wind zone with high pressure at 30° N. The average annual solar radiation is 5.28 kWh m^{-2} day^{-1}, exhibiting a seasonal pattern with the highest values occurring in the summer months. The annual mean temperature is 21.9 °C. The average temperatures of the coldest month (January) is 12.9 °C while the hottest month (July) is 28.8 °C, resulting in a seasonal monthly pattern similar to that of solar radiation. The temperatures show a variation over the years with an apparent increase in the frequency of extreme cold events during the winter and extreme hot events during the summer. These results suggest that the winters are becoming colder while the summer months are becoming warmer. This annual variability is associated with the North Atlantic Oscillation (NAO). The annual averages of potential evapotranspiration and annual precipitation are 2602 mm $year^{-1}$ and 211 mm $year^{-1}$, respectively, suggesting that the average annual rainfall only represents only 8% of the annual water for evapotranspiration demand. The annual precipitation also shows high variability over the years as a consequence of El Niño, NAO, and tropical cyclones. The models under global climate change predict that the climate of CCB has a trend of becoming drier and warmer with a high frequency of extreme climatic events, resulting in a more extreme climate.

C. Montiel-González · F. García-Oliva (✉)
Instituto de Investigaciones en Ecosistemas y Sustentabilidad, Universidad Nacional Autónoma de México, Morelia, Michoacán, Mexico
e-mail: fgarcia@cieco.unam.mx

F. Bautista
Centro de Investigaciones en Geografía Ambiental, Universidad Nacional Autónoma de México, Morelia, Michoacán, Mexico

C. Delgado
Laboratorio Universitario de Geofísica Ambiental, Instituto de Geofísica. Universidad Nacional Autónoma de México, Morelia, Mexico

V. Souza et al. (eds.), *Cuatro Ciénegas Ecology, Natural History and Microbiology*, Cuatro Ciénegas Basin: An Endangered Hyperdiverse Oasis,
https://doi.org/10.1007/978-3-319-93423-5_3

Introduction

The deserts are the most distributed ecosystems in the world, because they cover about 40% of the Earth's surface (UNEP 1997). The water availability in these ecosystems is the most limiting factor for plant productivity (Williams 2014), because the annual precipitation represents only 65% of the evapotranspiration demand (FAO 2007).

Desert ecosystems comprise nearly 50% of Mexico (Rzedowski 2006). One of the most important deserts in North America is the Chihuahuan Desert, because it covers 36% of the North American dry lands and represents an important hot spot of biodiversity (Archer and Predick 2008). Within this desert, the Cuatro Ciénegas Basin (CCB) is considered the most important wetland of Mexico (Souza et al. 2011).

This chapter describes the factors that determine the CCB climate, the temporal variability of climate parameters, and the climate scenarios under global climate change. To meet these objectives, we used a temperature and precipitation dataset covering a period of 76 years (1941–2017) taken from the weather station within CCB (http://smn.cna.gob.mx/). The resulting data were processed with Clic-MD software (Bautista et al. 2016). The daily solar radiation data covering a period of 30 years were taken from dataset surface meteorology and solar energy (NASA 2018).

Temporal Pattern of Solar Radiation and Temperature in CCB

The average annual solar radiation within CCB is 5.28 kWh m^{-2} day^{-1}. Figure 3.1a shows monthly solar radiation reflecting its seasonal pattern, with the lowest values in December and January, while the highest values are between April and July. These data are a result of global net solar radiation variation (Bonan 2015).

The annual mean temperature (Tmean) is 21.9 °C. The average temperatures of the coldest month (January) is 12.9 °C and of the hottest month (July) is 28.8 °C, resulting in a seasonal monthly pattern similar to that of solar radiation (Fig. 3.2). However, the differences between the annual average maximum temperature (Tmax) and minimum temperature (Tmin) is 14.6 °C. The winter months have a Tmax above 20 °C (December and January), while the hottest months in the summer are above 34.7 °C (June to August; Fig. 3.2). Similarly, the Tmin has a seasonal monthly pattern wherein the winter months are below 5.5 °C, and the summer months are above 22 °C (Fig. 3.2). The monthly differences between Tmax and Tmin ranged between 12.0 °C and 16.6 °C with the highest values occurring between December and April. These data suggest that the winter months have a higher thermal oscillation than the summer months.

However, the temperatures show an important variation over the years (Fig. 3.3); we observed that the Tmean and Tmin of the summer months increased up to 2 °C from 1941 to 2013. Additionally, we reported that the frequency of extreme events of

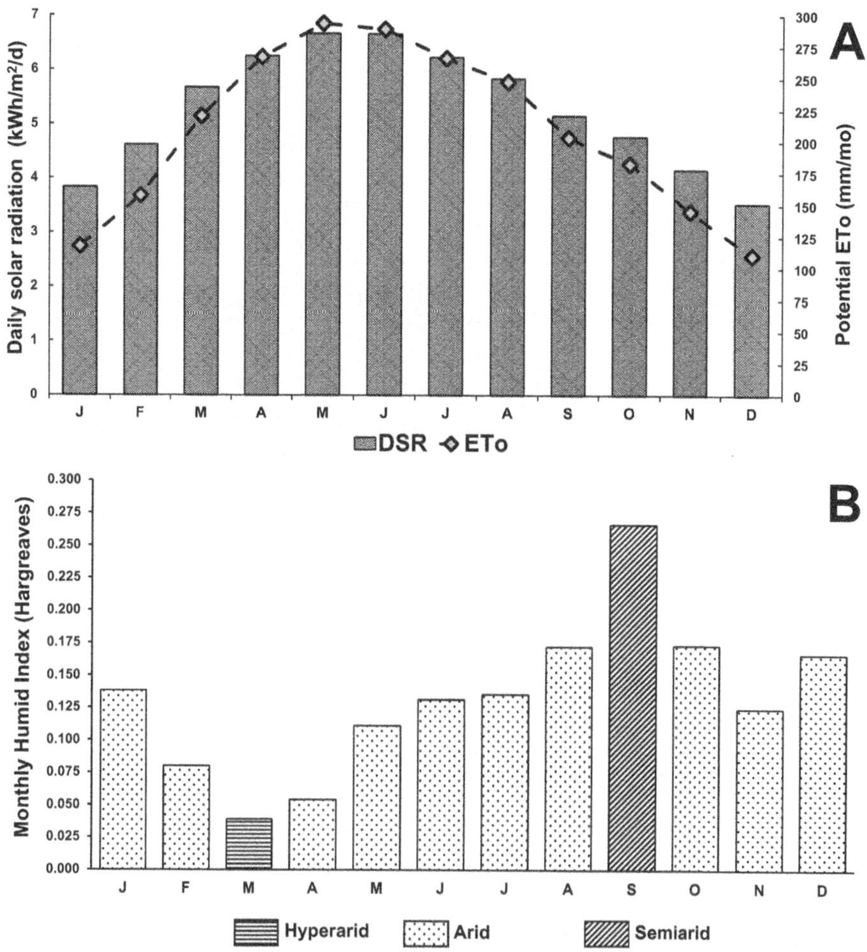

Fig. 3.1 The monthly average of (**a**) the daily solar radiation (kWh m^{-2} day^{-1}) and potential evapotranspiration (mm m^{-1}) from CCB (Dataset surface meteorology and solar energy, NASA 2018) and (**b**) Monthly Humid Index calculated by Hargreaves (Dataset SMN years1941–2017; Clic-MD software)

Tmin during the winter and extreme events of Tmax during the summer increased in the last 30 years (Montiel et al. 2017). These results suggest that the winters are becoming colder, while the summer months were becoming warmer, leading to an increase in the frequency of cold waves and heat waves in the last 30 years in CCB. This annual variability is associated with the North Atlantic Oscillation (NAO).

The NAO is a variation of the air surface pressure from north to south—between the Icelandic Low and the Bermuda-Azores High—that affects the wind direction and location of cold storms in the North Atlantic during the winter months (Scaife et al. 2014). The NAO consists of two phases, a positive and a negative (Archer and Predick 2008). There is an intensification of the pressure differences in the positive

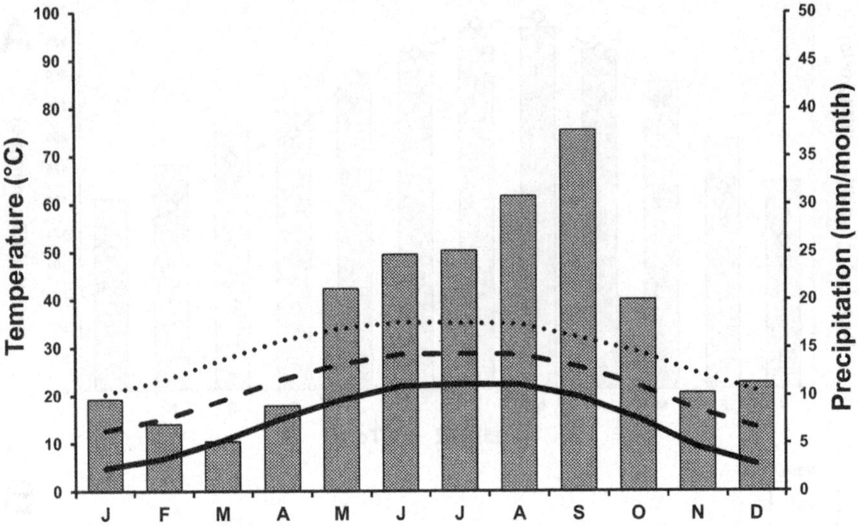

Fig. 3.2 Monthly average temperatures (Tmin, minimum; Tmean, mean; and Tmax, maximum values) and monthly average precipitation recorded from 1941 to 2017 (Dataset SMN; Clic-MD software)

NAO phase, because the Bermuda–Azores High pressure increases and the Icelandic Low pressure decreases. The pressure difference is reduced between both pressure centers in the negative NAO phase. In Northern Mexico, the positive NAO produces warmer winters while the negative phase produces extremely cold winters with frost and cold waves (Archer and Predick 2008).

Temporal Patterns of Evapotranspiration, Precipitation, and Aridity Index in the CCB

The annual averages of potential evapotranspiration and annual precipitation in the CCB are 2602 mm $year^{-1}$ and 211 mm $year^{-1}$, respectively, suggesting that the annual rainfall only represents about 8% of the annual water demand by evapotranspiration. Therefore, the aridity index proposed by UNEP (1997) is 0.08, corresponding to an arid climate.

The aridity condition of CCB is a consequence of its geographic location. The CCB is located at 26° 50′ 41″ N and is strongly affected by a divergent wind zone with high pressure at 30° N (Bonan 2015). This divergent wind zone is produced by descending cold and dry winds generating dry conditions on the land surface (Bonan 2015). This arid wind belt in the continental southern sites creates arid conditions mainly during the winter season (Ahrens 2013; Bonan 2015).

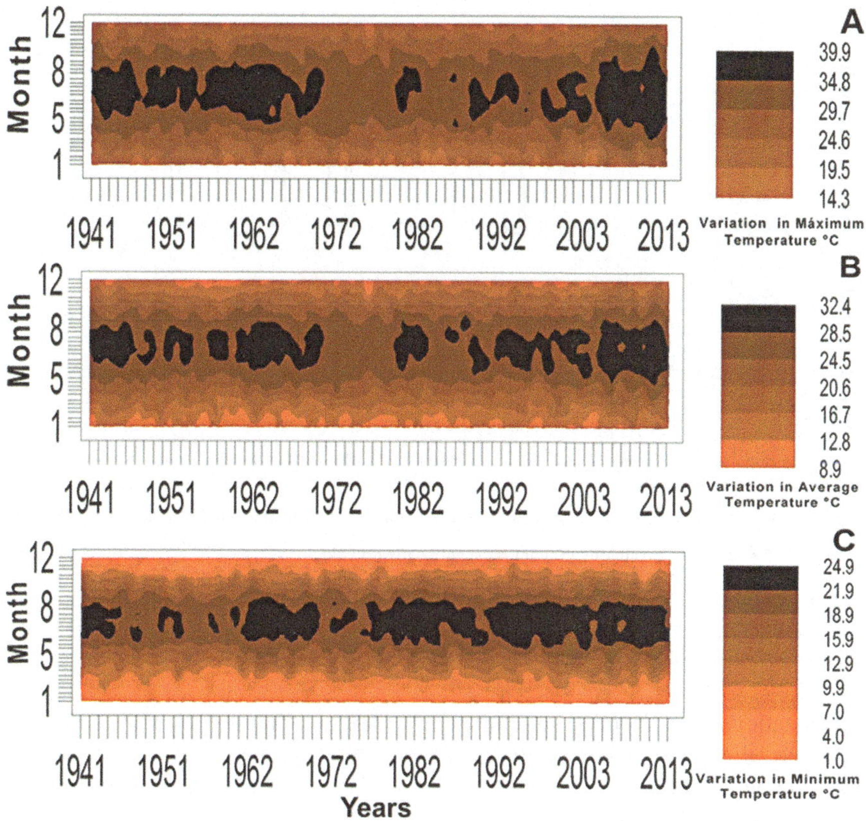

Fig. 3.3 The temporal temperature pattern in a 70-year period (1941–2013) taken from the weather station within CCB (http://smn.cna.gob.mx/): (**a**) maximum temperature (Tmax), (**b**) mean temperature (Tmean), and (**c**) minimum temperature (Tmin)

As a consequence of the monthly solar radiation pattern, the water demand for evapotranspiration is high during the months with elevated solar radiation (April to July; Fig. 3.1a), when the annual rainfall is concentrated. The summer months contribute to about 90% of the annual precipitation, while the winter months contribute to about 9% (Fig. 3.2). According to the monthly humid index calculated with the Hargreaves equation, most months have arid conditions except for September and March which have semiarid and hyperarid conditions, respectively (Fig. 3.1b).

Additionally, the annual precipitation exhibits substantial variation. Therefore, the standardized drought-humid index changes greatly over the years (Fig. 3.4). We found several years to be exceptionally humid or exceptionally arid. Additionally, the frequency and intensity of these extreme events apparently increased after 1985. The variability in the arid conditions between years can be explained in part by El Niño Southern Oscillation (ENSO) and La Niña and the North Atlantic Oscillation (NAO). During El Niño years, the climate in Northern Mexico shows an increase in

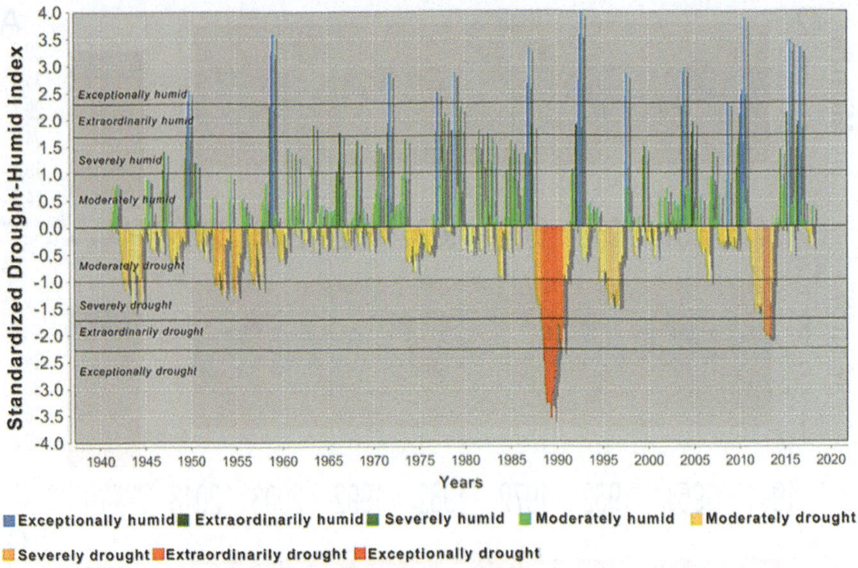

Fig. 3.4 Standardized drought-humid index. Positive values indicate standardized humid index, and the negative values indicate standardized drought index. The severe to exceptional events indicate a *Niño* or *Niña* year (Dataset SMN years1941–2017; Clic-MD software)

precipitation and a decrease in temperature, especially during the winter months. However, this phenomenon produces less precipitation and hotter temperatures in the summer months, thus creating drought and heat waves (CONAGUA 2017; Magaña et al. 1998; Vose et al. 2014). During La Niña years, the climate in the Northern Mexico shows an increase in precipitation during the summer months. (Magaña et al. 1998). The frequency and intensity of El Niño (negative values in Fig. 3.4) and La Niña (positive values in Fig. 3.4) have strongly influenced the extreme events of drought and storms in the last two decades. Additionally, the positive NAO phase has produced wetter winters than normal (Archer and Predick 2008).

The CCB climate type, according to Köopen and modified by García (1981), is BWhwx′(w)(e′) calculated from 30 years of climate records (between 1977 and 2013). This climate type is very dry and semi-warm with an annual average temperature of 21.9 °C. Thermal oscillation is very extreme. Summer rains contribute about 90% of the annual precipitation.

Climate Scenarios Under Global Climate Change

According to the IPCC (Stocker et al. 2013), the average global temperature increased 0.85 °C from 1880 to 2012, but global changes in precipitation show no clear temporal trend (Stocker et al. 2013). Similarly, heat waves have increased in intensity since the mid-twentieth century in most of the world, but the temporal pattern of torrential rains and drought remains unclear at the global level (Stocker et al. 2013). It is projected that under the worst scenario of greenhouse gas emissions (RCP8.5 model), the average temperature could increase by 3 or 4 °C, and the average annual precipitation could decrease by 10–20% by 2099 in the arid zones of Northern Mexico (SEMARNAT-INECC 2016).

Likewise, extreme weather events are expected to be more frequent and more extreme (IPCC 2012). An increase in drought, heat wave, and frost events is projected for 2100 in the arid zones of Mexico (SEMARNAT-INECC 2016). These results suggest that the climate of CCB will become drier and warmer with a high frequency of extreme climatic events, resulting in an even more extreme climate.

Acknowledgments The authors thank the Servicio Meteorológico Nacional of México (SMN) and the Instituto Nacional de Investigaciones Forestales Agrícolas y Pecuarias (INIFAP) for providing meteorological data. The authors thank Skiu (Scientific Knowledge In Use www.actswith-science.com) for granting a Clic-MD software license for data analysis.

References

Ahrens CD (2013) Meteorology today: an introduction to weather, climate, and the environment. Thomson/Brooks/Cole, Belmont

Archer SR, Predick KI (2008) Climate change and ecosystems of the southwestern United States. Rangelands 30:23–28

Bautista F, Pacheco A, Bautista-Hernández DA (2016) Climate change analysis with monthly data (Clic-MD). Skiu. México, CDMX, 57 p

Bonan G (2015) Ecological climatology: concepts and applications. Cambridge University Press, Cambridge

CONAGUA (2017) Impactos del ENOS en México. http://smn.cna.gob.mx/es/8-smn-general/92-enos. Accessed Oct 2017

FAO (2007) Carbon sequestration in dryland soils. World Soil Resources Reports. No. 102. World Soil Resources Reports Rome

García E (1981) Modificación al sistema de clasificación climática de Köppen. Instituto de Geografía. Universidad Nacional Autónoma de México, Mexico City

IPCC (2012) Summary for policymakers. In: Field CB, Barros V, Stocker TF, Qin D, Dokken DJ, Ebi KL, Mastrandrea MD, Mach KJ, Plattner GK, Allen SK, Tignor M, Midgley PM (eds) Managing the risks of extreme events and disasters to advance climate change adaptation a special report of working groups I and II of the intergovernmental panel on climate change. Cambridge University Press, Cambridge, pp 1–19

Magaña V, Pérez JL, Conde C (1998) El fenómeno de El Niño y la oscilación del sur. Sus impactos en México. Ciencias 51:14–18

Montiel C et al (2017) The response of soil microbial communities to variation in annual precipitation depends on soil nutritional status in an oligotrophic desert. Peer J 5:e4007. https://doi.org/10.7717/peerj.4007

NASA (2018) Surface meteorology and solar energy. https://eosweb.larc.nasa.gov/sse/RETScreen/. Accessed Feb 2018

Rzedowski J (2006) Vegetación de México. Limusa, Mexico City

Scaife AA, Arribas A, Blockley E, Brookshaw A, Clark RT, Dunstone N, Eade R, Fereday D, Folland CK, Gordon M, Hermanson L, Knight JR, Lea DJ, MacLachlan C, Maidens A, Martin

M, Peterson AK, Smith D, Vellinga M, Wallace E, Waters J, Williams A (2014) Skillful long range prediction of European and north American winters. Geophys Res Lett 41:2514–2519
SEMARNAT-INECC (2016) Mexico's climate change mid-century strategy. Ministry of Environment and Natural Resources (SEMARNAT) and National Institute of Ecology and Climate Change (INECC), Mexico City
Souza V, Siefert JL, Escalante AE, Elser JJ, Eguiarte LE (2011) The Cuatro Ciénegas Basin in Coahuila, Mexico: an Astrobiological Precambrian Park. Astrobiology 12:641–647
Stocker TF, Qin D, Plattner GK, Tignor M, Allen SK, Boschung J, Nauels A, Xia Y, Bex V, Midgley PM (eds) (2013) The physical science basis. Contribution of working group I to the fifth assessment report of the intergovernmental panel on climate change. Cambridge University Press, Cambridge
UNEP (1997) In: Middleton N, Thomas DSG (eds) World atlas of desertification. UNEP: London
Vose RS, Applequist S, Bourassa MA, Pryor SC, Barthelmie RJ, Blanton B, Bromirski PD, Brooks HE, DeGaetano AT, Dole RM, Easterling DR, Jensen RE, Karl TR, Katz RW, Klink K, Kruk MC, Kunkel KE, MacCracken MC, Peterson TC, Shein K, Thomas BR, Walsh JE, Wang XL, Wehner MF, Wuebbles DJ, Young RS (2014) Monitoring and understanding changes in extremes: extratropical storms, winds, and waves. Bull Am Meteorol Soc 95:377–386
Williams M (2014) Climate change in deserts: past, present and future. Cambridge University Press, Cambridge

Chapter 4
Hyperdiverse Viral Communities in an Oligotrophic Oasis (Cuatro Ciénegas): Marine Affinities and Microgeographic Differentiation

P. Isa, B. Taboada, A. L. Gutiérrez, P. Chávez, R. M. del Ángel, J. E. Ludert, A. C. Espinosa, L. E. Eguiarte, E. Garrido, S. López, V. Souza, and C. F. Arias

Abstract The Cuatro Ciénegas Basin (CCB) is a biologically exceptional place. In general, CCB aquatic systems are extremely oligotrophic, but paradoxically they are extraordinarily biodiverse, and viruses are not an exception. Previous studies had shown that two stable ecosystems within CCB had very different virus; one of them, Pozas Azules, presented divergent virus with a marine affiliation. In this chapter, we compare an earlier study with the virome of three sites at CCB, sampled in 2014, including fish guts from one of the sites. The idea was to test if the virus follows, like a mirror image, the diversity of a site, by showing us the predators vs their prey.

In order to do so, we included all type of virus in this analysis and had a very good coverage in the sequencing. For every sample, more than half of the reads did not match any known sequence deposited in databases. We found a very high virus diversity in all samples. We detected 1691 different viral species, belonging to 170 genera, classified in 40 viral families. It is important to mention that these viruses represent 38.5% of viral families characterized today. At genus level, 33.7% of known viral genera were detected in CCB in the last sampling, with more than half species described (53.8%). The relation between the viral community patterns

---

P. Isa and B. Taboada contributed equally

P. Isa · B. Taboada · S. López · C. F. Arias
Instituto de Biotecnología, Universidad Nacional Autónoma de México, Cuernavaca, Morelos, Mexico

A. L. Gutiérrez · P. Chávez · R. M. del Ángel · J. E. Ludert · E. Garrido
Centro de Investigación y de Estudios Avanzados, Instituto Politécnico Nacional, Ciudad de México, Mexico

A. C. Espinosa · L. E. Eguiarte · V. Souza (✉)
Instituto de Ecología, Universidad Nacional Autónoma de México, Ciudad de México, Mexico
e-mail: souza@unam.mx

V. Souza et al. (eds.), *Cuatro Ciénegas Ecology, Natural History and Microbiology*, Cuatro Ciénegas Basin: An Endangered Hyperdiverse Oasis,
https://doi.org/10.1007/978-3-319-93423-5_4

within and among the sampling aquatic systems in CCB was tested using a canonical analysis of principal coordinates and a nonmetric multidimensional scaling (NMDS) on a dissimilarity matrix from species abundance in each sample. The analysis showed three separated clusters associated to the sampling sites, being Pozas Rojas very different to Churince and La Becerra, which makes sense since they are very close geographically; a third cluster was the fish samples since by their diet the fish gut enriched for invertebrates and plants virome. The virome diversity of the CCB samples indicated a high diversity compared to other water samples, either from freshwater or ocean samples, suggesting that CCB water systems are among the highest virus biodiversity estimates recorded.

Introduction

The Cuatro Ciénegas Basin (CCB) is a biologically exceptional place. In general, CCB aquatic systems are extremely oligotrophic, but paradoxically they are extraordinarily biodiverse. Located in the center of Chihuahuan desert, in the state of Coahuila, Mexico, CCB is a small intermountain valley with a basin floor ca. 740 m above sea level, surrounded by mountains more than 3500 m high. CCB comprises hundreds of "pozas"—lakes, ponds, and even rivers of variable sizes and characteristics—formed by seven main drainage systems (Minckley 1969). Indeed, CCB has been characterized as biologically hyperdiverse, because it has the greatest number of endemic species of any place in North America (Stein et al. 2000); it has a high biodiversity of plants, vertebrates, and invertebrates (Cole 1984; Lemos-Espinal and Smith 2016; Minckley 1969) as well as abundant living microbialites (including stromatolites, thrombolites, etc.) and microbial mats (Souza et al. 2006, 2007; Stein et al. 2000). All our studies show that in part the reason for microbial hyperdiversity is the local adaptation of each community; hence, we have a site with a very high beta-diversity (i.e., communities are very different among sites) in all its explored biomes (Bonilla-Rosso et al. 2012; Cerritos et al. 2011; Escalante et al. 2008; Moreno-Letelier et al. 2012; Souza et al. 2006; Espinosa-Asuar et al. 2015; Pajares et al. 2018). This high biodiversity is fascinating and apparently a paradox, since as mentioned above, CCB is very poor in nutrients (Lopez-Lozano et al. 2012; Tapia-Torres et al. 2015) and also has low primary productivity (Tapia-Torres et al. 2015). Nevertheless, it is relevant to mention that CCB is now extremely endangered because of recent excessive water exploitation in the area (Souza et al. 2006, 2007).

Previous studies of CCB bacteria have reported signals of marine ancestry (i.e., the CCB lineages are more similar to marine strains than to any other samples), including studies using cultured bacteria (Alcaraz et al. 2008; Cerritos et al. 2011; Moreno-Letelier et al. 2012), environmental 16S rDNA clones (Souza et al. 2006), or metagenomic data (Peimbert et al. 2012). Nevertheless, it is thought that the diversity of CCB is a result of local adaptations and consequently has high divergence as has been detected when its communities have been sequenced or when particular bacterial strains are compared with samples in other localities in the planet, and also a high beta-diversity among sites within CCB.

Viruses, by definition, are obligatory intracellular parasites, and they have developed the capacity to infect all forms of life, from unicellular to complex organisms. They represent the most abundant and diverse biological entities in all Earth ecosystems, including extreme environments as volcanic hot springs (Breitbart et al. 2004), oceanic basement (Nigro et al. 2017), or very isolated water holes (Anesio et al. 2007). Viruses affect distinct parts of the life cycle of the organism they infect and have important impacts on their population dynamics and microevolution. By these means, viruses modify and model the ecology and even the evolution of complete ecosystems (Fancello et al. 2013): we can mention marine cyanophages (viruses specialized in cyanobacteria) that are very important in sea ecology by killing the main primary producers and thus modifying complete ecological cycles, including nitrogen, carbon (in particular photosynthetic patterns), and phosphorus, among other critical processes (Clokie and Mann 2006; Lu et al. 2001).

Despite their usual high prevalence and divergence, there is still scant information regarding viruses in CCB. Desnues et al. in 2008 first analyzed the biodiversity and biogeography of phages in modern thrombolites and stromatolites in two aquatic systems (Río Mezquites and Pozas Azules) using 454 technologies. Both systems were different, while Río Mezquites thrombolite was not very diverse and dominated by cyanophages. Pozas Azules stromatolite showed a very high diversity of virus and genetic similarities in their phage communities between CCB and marine samples (i.e., Sargasso Sea and Gulf of Mexico).

Herein we describe the diversity of viruses present in three aquatic systems in CCB sampled in 2014, using massive next-generation sequencing (NGS) strategies and expanding the results and discussion of Taboada et al. (2018). One aquatic system, La Becerra, has been used for swimming and recreation for decades. This site is now under recovery since 2009, when it was closed to the public. Another system, Pozas Rojas, has little human disturbance but experienced a strong perturbation in 2010 when a hurricane drastically increased the water level in these pozas. The Churince system has become increasingly dryer in recent years, apparently because of the intense use of the deep water that feeds this system (Souza et al. 2007). In addition, we studied the viruses from the intestinal contents of three fish species from Churince.

This study, as most other chapters in this book and in the book series, is part of the *General Inventory of Churince, a* project supported by the generous special WWF-FCS grant. This extensive project was an enormous effort to describe "as much as possible and as fast as possible" the Churince ecosystem before it completely dried because of the drainage of its water sources. We believe that by analyzing the virosphere (i.e., the inventory of all viruses present), we could explore the idea that viruses could represent the "mirror image" of their host biodiversity. The previous study of Desnues et al. (2008) described above studied the virome before the water levels in some systems in CCB (in particular Churince) became problematic. That study suggested that virus had followed the marine ancestry of their marine bacteria host, and as a result of their prolonged isolation, they are also mainly endemic to CCB, an observation that is very seldom reported for viruses. This chapter will show evidences of this claim.

Virus Diversity in CCB

To characterize viruses present in three sampled aquatic systems of CCB in 2014, genetic material extracted from water was subjected to next-generation technology (Illumina), allowing us to obtain millions of short reads. Additionally, to test viruses as "mirrors" of their host diversity and abundance, we also analyzed the genetic material of fish intestinal samples. The basic idea was that as fish eat different food, different viral compositions will be present, even when coexisting in the same poza. General materials and methods are to be consulted in Taboada et al. (2018).

For every sample, more than half of the reads did not match any known sequence deposited in databases (Fig. 4.1), confirming the presence of the so-called genetic dark matter, which could, in theory, represent many still unknown viruses, similarly to previous studies done with environmental samples (Brum et al. 2015).

We found a very high virus diversity in all samples. We detected 1691 different viral species, belonging to 170 genera, classified in 40 viral families. It is important to mention that these viruses represent 38.5% of viral families characterized today. At genus level, 33.7% of known viral genera were detected in CCB in the last sampling, with more than half of the species described (53.8%). These results confirmed a remarkable virus biodiversity in the sampled system of CCB. Among the reads and different types of viral genomes detected, the most prevalent corresponded to dsDNA viruses, followed by ssDNA, ssRNA, and dsRNA viruses (Tables 4.1 and 4.2). A considerable number of novel viruses were also observed, supporting the previously suggested idea of CCB being a "lost world" where ancient lineages persisted (Moreno-Letelier et al. 2012).

Most prevalent viral reads identified in the water samples were related to bacterial hosts, representing in average 71.6% of all viral reads. Phage genomes corresponding to six different dsDNA virus families were found; among them, members of the *Podoviridae, Myoviridae*, and *Siphoviridae* families were the most abundant. In fact, phages belonging to *Siphoviridae* and *Podoviridae* family were observed by electron microscopy (Fig. 4.2). One member of the ssDNA *Microviridae* family was also found, although with a lower frequency.

On the other hand, in all samples abundant reads that shared identity to the eukaryotic dsDNA viruses from the *Phycodnaviridae* (that infects algae) and of the *Iridoviridae* (that infects invertebrates and vertebrates) families were also present. Reads similar to viruses infecting ameba (from the *Mimiviridae* family) were also found in all samples. Interestingly, most of these viruses are novel members of the identified virus families, confirming the uniqueness of CCB.

Besides these major viral families, the remaining 28 families were sporadic. An interesting find was that all dominant phages in CCB have also been isolated from

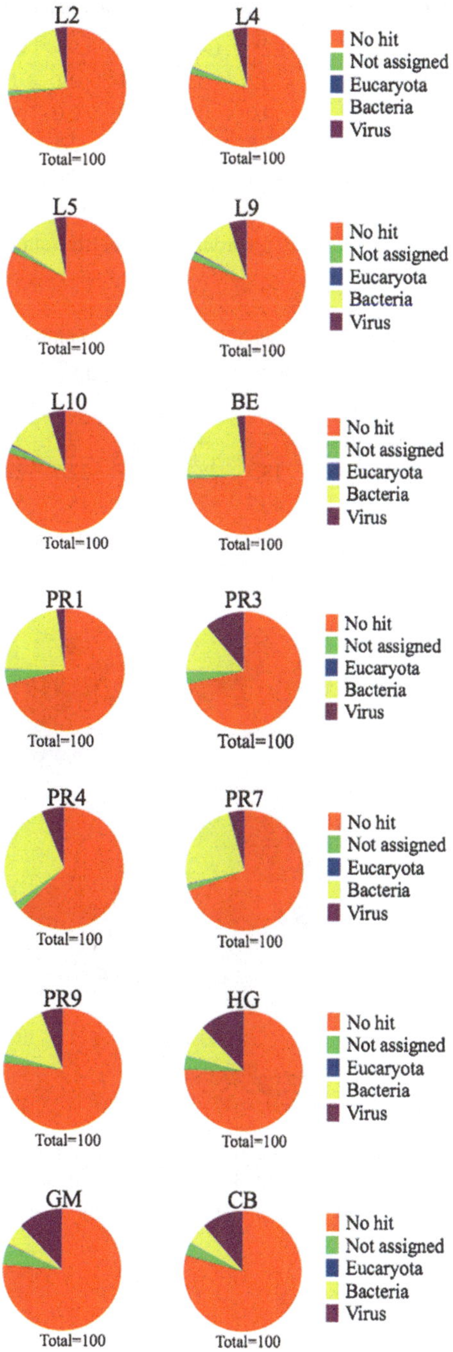

Fig. 4.1 Relative abundance of sequence reads based in MEGAN taxonomic assignment. Reads that did not have any identity to known sequences in the databases were placed into the "No hit" category. The reads assigned to viral taxa that do not meet selected parameters were placed into "Not assigned" category. L2 to L10 are samples from Churince; BE is the sample from La Becerra and PR1 to PR9 are samples from Pozas Rojas. Fish samples were marked as *Hemichromis guttatus* (HG), *Gambusia marshi* (GM), and *Cyprinodon bifasciatus* (CB)

Table 4.1 Percentage of reads assigned to the type of virus genome

	L2[a]	L4	L5	L9	L10	BE	PR1	PR3	PR4	PR7	PR9
ssRNA viruses	0.2	0.2	0.1	0.1	0.0	6.7	0.0	0.0	0.4	0.2	0.1
dsRNA viruses	0.0	0.0	0.0	0.0	0.0	0.6	0.0	0.0	0.0	0.0	0.0
ssDNA viruses	7.8	4.8	4.9	2.8	2.8	9.5	0.2	0.2	3.7	9.5	0.5
dsDNA viruses	47.4	58.3	47.0	58.8	56.4	66.1	50.2	68.1	65.1	49.0	61.9
Unknown	44.5	36.7	48.0	38.2	40.7	17.2	49.6	31.6	30.9	41.3	37.5

[a]Samples L2 to L10 are from Churince, BE from La Becerra, and PR2 to PR9 from Pozas Rojas

Table 4.2 Percentage of identified virus families assigned by their genome types

	L2[a]	L4	L5	L9	L10	BE	PR1	PR3	PR4	PR7	PR9
ssRNA viruses	33.3	28.6	30.4	22.7	17.6	46.7	14.3	23.5	21.7	11.8	15.8
dsRNA viruses	7.4	4.8	0.0	9.1	0.0	6.7	0.0	0.0	4.3	5.9	5.3
ssDNA viruses	18.5	23.8	21.7	22.7	29.4	16.7	21.4	29.4	21.7	29.4	26.3
dsDNA viruses	40.7	42.9	47.8	45.5	52.9	30.0	64.3	47.1	52.2	52.9	52.6

[a]Samples L2 to L10 are from Churince, BE from La Becerra, and PR2 to PR9 from Pozas Rojas

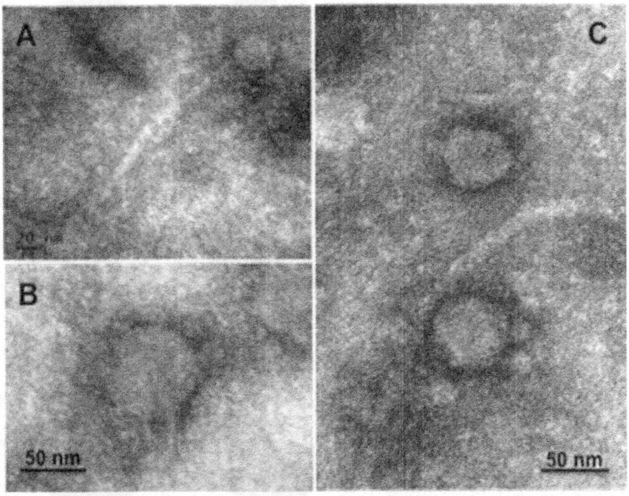

Fig. 4.2 Electron micrograph of virus particles observed in CCB samples. (**a**) Non-enveloped tailed phages compatible with members of the *Siphoviridae* family; (**b**, **c**) hexagonal head and short-tailed phages compatible with members of the *Podoviridae* family

seawater, supporting the idea that bacteria related to marine lineages are abundant in both CCB systems (Taboada et al. 2018).

Microgeographic Structure of CCB Viruses

When analyzing ecosystem metagenomics data, it is crucial to assess the diversity of the viral species of interest. We applied different methods to measure richness and evenness of the viral data.

Analyses by the nonparametric richness estimator of Chao indicated that each aquatic system (i.e., Churince, La Becerra, and Pozas Rojas) was well sampled, as it had reached the expected viral diversity by region. Overall, the Shannon diversity indexes, which take into account diversity and equitability of a sample, show that all metagenomes have high richness. Additionally, high evenness values in the Pielou scale of 0.77 to 0.89 were obtained within the aquatic systems, while for fish samples, they were close to 1 (0.91 to 0.93), indicating that the different viral genotypes are close to being equally abundant (Taboada et al. 2018), suggesting a "mirror image" of equitative distribution of hosts.

The relation between the viral community patterns within and among the sampling aquatic systems in CCB was tested using a canonical analysis of principal coordinates and a nonmetric multidimensional scaling (NMDS) on a dissimilarity matrix from species abundance in each sample. The analysis clearly showed three separated geographical clusters (Fig. 4.3); the first one, that is very different form the other two clusters, contained viruses identified in intestinal contents of three species of fish from Churince. A second cluster includes the water samples from Churince and La Becerra systems, which are geographically close (less than 5 km), and in both sulfate salts dominate the chemistry of water. A third cluster includes all the five samples from Pozas Rojas, on the other side of the valley, where other salts (mainly calcium carbonate) dominate the chemistry of the water (Peimbert et al. 2012; Rebollar et al. 2012).

Viral Diversity Is Related to the Ecosystem Diversity and Structure

A high diversity of viral genomes is present in the very endangered and disturbed aquatic system of Churince, as well as in La Becerra and Pozas Rojas, two systems under recovery after different types of perturbation. DNA and RNA viruses are present in the various analyzed systems (Desnues et al. 2008; Taboada et al. 2018), but with a vast dominance of DNA viruses while RNA viruses are rare. This inequality can be the result of (1) a real lower abundance of RNA viruses, (2) the smaller

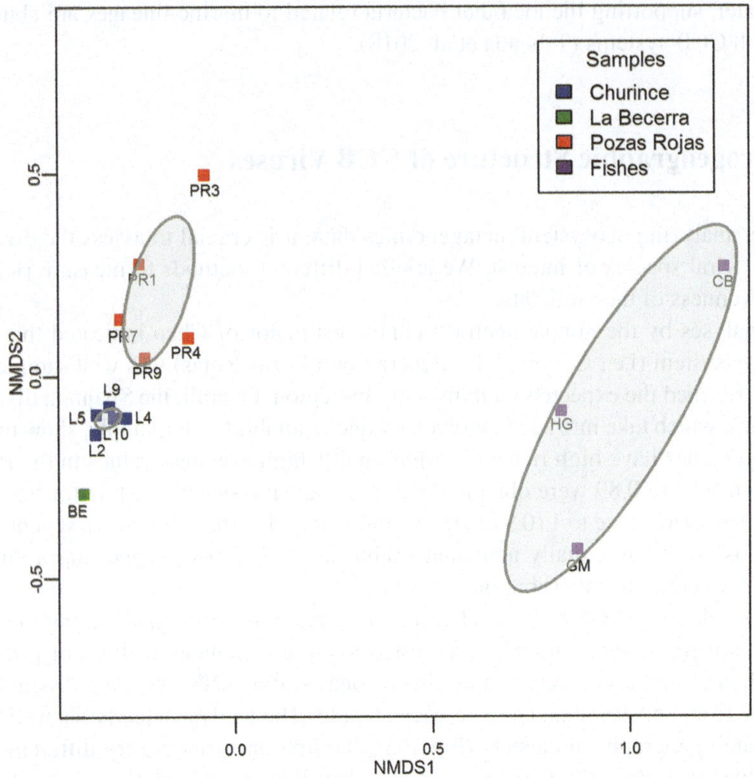

Fig. 4.3 Nonmetric multidimensional scaling (NMDS) of viral communities at the species level. Each point corresponds to a sample, and ellipses represent the standard deviation of the weighted averages of the drainage systems and fish intestinal contents. Ellipses were calculated using the ordiellipse function of R package "vegan" at 95% confidence. Samples are labeled as mentioned in Fig. 4.1

sizes of their genomes, or (3) RNA degrading easier and faster than DNA. Our finding is in accordance with the few studies that have addressed the lower abundances of RNA viruses in the sea, in comparison to DNA viruses (Brum et al. 2015).

A very high percentage of the sequences identified in this study (63.1–82.2%) are new to databases, and we reported it as unclassifiable "dark viral matter" (Fig. 4.1; see also Taboada et al. 2018). Our estimates are similar to the values reported in an extensive virus survey of the seas (TARA), where the amount of what is considered as an unknown (i.e., dark matter) was between 63 and 93% (Brum et al. 2015). Still uncharacterized viruses may be represented in the unknown sequences from CCB and elsewhere.

The most common viruses found were bacteriophages and viruses that infect invertebrates and plants, followed by sequence reads corresponding to eukaryotic dsDNA virus families that include members of the *Phycodnaviridae*, which infect algae (Zhang et al. 2015). This was surprising, given the oligotrophic nature of the

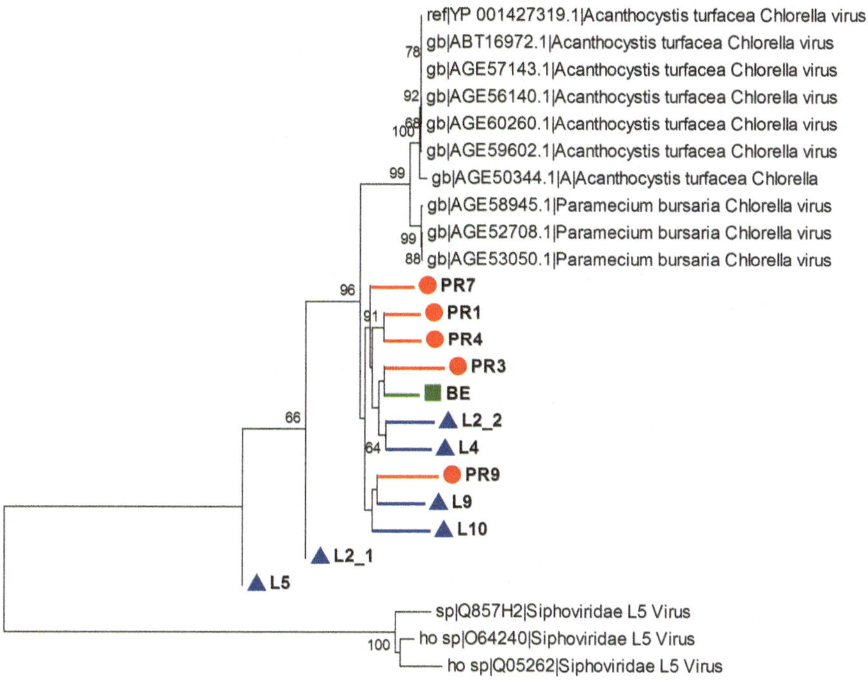

Fig. 4.4 Maximum likelihood tree of the ribonucleoside-triphosphate reductase protein of the *Phycodnaviridae* family. A total of ten different *Phycodnaviridae* sequences were used as reference. Three *Siphoviridae* were also included since they were other homologous sequences from GenBanl according to our analysis. Branches in the tree are colored by the following code: samples from Churince in blue, from Pozas Rojas in red, and from La Becerra in green. Bootstrap values are indicated for each node having at least a 50% support

ponds, where no visible algae are observed, although indeed microalgae are abundant and very diverse, at least in the Churince system. Viruses that infect free-living ameba, from the family *Mimiviridae*, were also found in all samples, suggesting that the protist biodiversity is also high, an observation confirmed by studies using 18S rDNA libraries.

It is important to note that the amino acid identities to known viral proteins were less than 60% in almost all cases, suggesting that maybe most of the viruses in CCB are novel members of the identified virus families. For example, the phylogeny analysis of the ribonucleoside-triphosphate reductase sequence of dsDNA viruses of the *Phycodnaviridae* family showed that none of the sequences obtained were closely related to any previously known clade in the world (Fig. 4.4). The viruses found in our study of CCB are mostly different from previously known viruses and highly diverse, confirming the initial observations by Desnues et al. (2008). These patterns are probably related to the endemicity, uniqueness, and hyperdiversity of the bacterial hosts and other organisms in CCB.

With respect to the abundance of virus species, it was interesting to find that, in the more stable La Becerra system, there were no viral species overrepresented, while both of the more fluctuant systems, Churince and Pozas Rojas, have dominant but different virus species, as illustrated by their distinct "clouds" in the NMDS analysis (Fig. 4.3). As expected, given its stability, the highest diversity was found in La Becerra water. Churince had a high diversity, similar to that observed in the less perturbed system of Pozas Rojas (Taboada et al. 2018). These systems differ from the samples analyzed by Desnues et al. (2008), as they sampled two other systems (Rio Mezquites and Pozas Azules) that represent two of the most stable systems within CCB.

According to the NMDS (Fig. 4.3), each fish species had different intestinal content viromes. *Cyprinodon bifasciatus* had the more differentiated virome, while *Gambusia marshi* and *Hemichromis guttatus* were more similar; this may be due to a lower virus diversity in *C. bifasciatus*, related to its more restricted diet and niche, as described recently by Rodríguez-Correa et al. (2017).

Comparison Between CCB Viromes and Other Viral Metagenomics Datasets

The taxonomic diversity of CCB was compared to the available taxonomic annotation datasets reported in different metagenomic analyses of virus communities, including the already described study of Desnues et al. (2008) of CCB microbialites, viromes from sea samples (Angly et al. 2006; Hurwitz and Sullivan 2013; Williamson et al. 2012), form freshwater samples (Djikeng et al. 2009; Fancello et al. 2013; Watkins et al. 2016), data from the MetaVir database (http://metavir-meb.univ-bpclermont.fr/), as described in Taboada et al. (2018). The virome diversity of the CCB samples indicated a high diversity compared to other water samples, either from the sea or from freshwater or ocean samples, suggesting that CCB water systems are among the highest virus biodiversity estimates recorded.

In particular, marine viruses have been studied in a more systematic and careful way than in other environs. For both diversity and abundance of marine viromes, host abundances and temperature are more relevant than geography to determine the viral communities (Brum et al. 2015; Thingstad 2000). Similar conclusions (i.e., that ecological conditions are more relevant than geographic distance) were observed in the viruses' communities in North American lakes by Djikeng et al. (2009) and by Watkins et al. (2016) and by Fancello et al. (2013) in ponds in Mauritania. In contrast, in the spindle-shaped viruses associated with the thermoacidophilic archaea *Sulfolobus islandicus*, a clear geographic pattern was found: samples from within a place were more related to the samples from that site and nearby sites than samples from faraway sites, i.e., US samples are more similar among them, Kamchatka samples are more similar among them and to the Japan sample, while Iceland samples are intermediate (Held and Whitaker 2009). Desnues et al. (2008) and Taboada et al. (2018) found a similar geographic association in the viral communities in CCB but at microgeographic scales, not at the worldwide scale of the Held and Whitaker (2009) study, which adjust also to the microgeographic differentiation patterns and strong biogeography of the bacteria in CCB (Escalante et al. 2008; Espinosa-Asuar et al. 2015; Pajares et al. 2018).

Models to Explaining the High Biodiversity and Microgeographic Patterns in CCB

As a result of the low concentrations of nutrients, in particular of phosphorus and nitrogen, one could expect that the viral communities in CCB have a strong local "microbial market," by increasing coexistence and complexity of the food webs (Ponce-Soto et al. 2015). A "kill the winner" model, where viruses concentrate their attacks in the most common strains in a given moment, further allows less abundant hosts to become more abundant, increasing in consequence both total virus and host densities and biodiversity (Cerritos et al. 2011; Escalante et al. 2008; Peimbert et al. 2012; Souza et al. 2006). Both processes could generate different local communities of viruses and bacteria, producing a pattern of strong microgeographic differentiation and high beta-diversity (Escalante et al. 2008; Espinosa-Asuar et al. 2015; Pajares et al. 2018). This is a likely scenario for CCB; hence, our "mirror image" hypothesis seems to be supported by both data and theory. We believe that viruses in CCB are following their host closely and have coevolved together, thus explaining why both bacteria and viruses have kept their ancient marine affiliations (Alcaraz et al. 2008; Cerritos et al. 2011; Moreno-Letelier et al. 2012). Direct experimental settings to "trap" such virus have been till now ineffective, given the very large diversity of potential hosts and the very small proportion of phyla that we can cultivate. However, molecular technology and nanotools are advancing so fast that maybe in the future we can design experiments where an ample range of interactions between prey and predators are measured, confirming, or not, the "kill the winner" idea.

Acknowledgments We thank the Instituto de Biotecnología-UNAM for granting access to its computer cluster, Jerome Verleyen and Juan Manuel Hurtado for their computational support, Laura Espinosa Asuar and Erika Aguirre Planter for technical assistance, the Protected Area APFF-Cuatro Cienegas Ciénegas for their logistics and facilitation, Arturo Gonzales from Museo del Desierto de Saltillo for their permit to work in La Becerra, the people of Ejido del Venado for the permit to work in Pozas Rojas, the high school CBTA22 for their molecular lab facilities, and the work of the high school students during the processing of the samples. This chapter was written during a sabbatical leave of VS in the Department of Ecology, Evolution, and Behavior, University of Minnesota, in Dr. Michael Travisano's laboratory, and of LEE, in the Department of Plant and Microbial Biology, University of Minnesota, in Dr. Peter Tiffin's laboratory, both supported by the program PASPA-DGAPA, UNAM.

References

Alcaraz LD, Olmedo G et al (2008) The genome of *Bacillus coahuilensis* reveals adaptations essential for survival in the relic of an ancient marine environment. Proc Natl Acad Sci U S A 105:5803–5808

Anesio AM, Mindl B et al (2007) Viral dynamics in cryoconite holes on a high Arctic glacier (Svalbard). J Geophys Res 112:04S31

Angly FE, Felts B et al (2006) The marine viromes of four oceanic regions. PLoS Biol 4:e368

Bonilla-Rosso G, Peimbert M et al (2012) Comparative metagenomics of two microbial mats at Cuatro Ciénegas Basin II: community structure and composition in oligotrophic environments. Astrobiology 12:15

Breitbart M, Wegley L et al (2004) Phage community dynamics in hot springs. Appl Environ Microbiol 70:1633–1640

Brum JR, Ignacio-Espinoza JC et al (2015) Ocean plankton. Patterns and ecological drivers of ocean viral communities. Science 348:1261498

Cerritos R, Eguiarte LE et al (2011) Diversity of culturable thermo-resistant aquatic bacteria along an environmental gradient in Cuatro Cienegas, Coahuila, Mexico. Antonie Van Leeuwenhoek 99:303–318

Clokie MR, Mann NH (2006) Marine cyanophages and light. Environ Microbiol 8:2074–2082

Cole GA (1984) Crustacea from the Bolson of Cuatro Cienegas, Coahuila, Mexico. J Arizona-Nevada Acad Sci 19:10

Desnues C, Rodriguez-Brito B et al (2008) Biodiversity and biogeography of phages in modern stromatolites and thrombolites. Nature 452:340–343

Djikeng A, Kuzmickas R et al (2009) Metagenomic analysis of RNA viruses in a fresh water lake. PLoS One 4:e7264

Escalante AE, Eguiarte LE et al (2008) Diversity of aquatic prokaryotic communities in the Cuatro Cienegas basin. FEMS Microbiol Ecol 65:50–60

Espinosa-Asuar L et al (2015) Aquatic bacterial assemblage structure in Pozas Azules, Cuatro Ciénegas Basin in México: deterministic vs. stochastic processes. Int Microbiol 18:105. https://doi.org/10.2436/20.1501.01.240

Fancello L, Trape S et al (2013) Viruses in the desert: a metagenomic survey of viral communities in four perennial ponds of the Mauritanian Sahara. ISME J 7:359–369

Held NL, Whitaker RJ (2009) Viral biogeography revealed by signatures in *Sulfolobus islandicus* genomes. Environ Microbiol 11:457–466

Hurwitz BL, Sullivan MB (2013) The Pacific Ocean virome (POV): a marine viral metagenomic dataset and associated protein clusters for quantitative viral ecology. PLoS One 8:e57355

Lemos-Espinal JA, Smith GR (2016) Amphibians and reptiles of the state of Coahuila, Mexico, with comparison with adjoining states. ZooKeys 593:117–137

Lopez-Lozano NE, Eguiarte LE et al (2012) Bacterial communities and the nitrogen cycle in the gypsum soils of Cuatro Cienegas Basin, coahuila: a Mars analogue. Astrobiology 12:699–709

Lu J, Chen F et al (2001) Distribution, isolation, host specificity, and diversity of cyanophages infecting marine Synechococcus spp. in river estuaries. Appl Environ Microbiol 67:3285–3290

Minckley W (1969) Environments of the Bolson of Cuatro Cienegas, Coahuila, Mexico. Sci Ser 2:65

Moreno-Letelier A, Olmedo-Alvarez G et al (2012) Divergence and phylogeny of Firmicutes from the Cuatro Cienegas Basin, Mexico: a window to an ancient ocean. Astrobiology 12:674–684

Nigro OD, Jungbluth SP et al (2017) Viruses in the oceanic basement. MBio 8:e02129–e02116

Pajares S et al (2018) Small-scale spatial heterogeneity in arid soils: microbial diversity and soil nutrients evidence. PeerJ 4:e2459. https://doi.org/10.7717/peerJ.2459

Peimbert M, Alcaraz LD et al (2012) Comparative metagenomics of two microbial mats at Cuatro Ciénegas Basin I: ancient lessons on how to cope with an environment under severe nutrient stress. Astrobiology 12:648–658

Ponce-Soto GY, Aguirre-von-Wobeser E et al (2015) Enrichment experiment changes microbial interactions in an ultra-oligotrophic environment. Front Microbiol 6:246

Rebollar EA, Avitia M et al (2012) Water-sediment niche differentiation in ancient marine lineages of *Exiguobacterium* endemic to the Cuatro Cienegas Basin. Environ Microbiol 14:2323–2333

Rodríguez-Correa H, González-Rodríguez A et al (2017) Perspectives of molecular ecology in a megadiverse country. Rev Mex Biodiversidad 81:10

Souza V, Espinosa-Asuar L et al (2006) An endangered oasis of aquatic microbial biodiversity in the Chihuahuan desert. Proc Natl Acad Sci U S A 103:6565–6570

Souza V, Falcón LI et al (2007) Protecting a window into the ancient earth: Towards a precambrian park at Cuatro Cienegas, Mexico. Citizen's Page, Evolutionary Ecology Research

Stein BA, Kutner LS et al (2000) Precious heritage: the status of biodiversity in the United States. Oxford University Press, New York

Taboada B, Isa P et al (2018) The viral structure of the Cuatro Ciénegas Basin, a unique oasis in Northern Mexico, reveals a highly diverse population at a small geographic scale. Appl Environ Microbiol 84:e00465

Tapia-Torres Y, Elser JJ et al (2015) Ecoenzymatic stoichiometry at the extremes: how microbes cope in an ultra-oligotrophic desert soil. Soil Biol Biochem 87(9)

Thingstad TF (2000) Elements of a theory for the mechanisms controlling abundance, diversity, and biogeochemical role of lytic bacterial viruses in aquatic systems. Limnol Oceanogr 45:8

Watkins SC, Kuehnle N et al (2016) Assessment of a metaviromic dataset generated from near-shore Lake Michigan. Mar Freshw Res 67:8

Williamson SJ, Allen LZ et al (2012) Metagenomic exploration of viruses throughout the Indian Ocean. PLoS One 7:e42047

Zhang W, Zhou J et al (2015) Four novel algal virus genomes discovered from Yellowstone Lake metagenomes. Sci Rep 5:15131

Chapter 5
How Divergent Is the Cuatro Ciénegas Oasis? Genomic Studies of Microbial Populations and Niche Differentiation

Zulema Gomez-Lunar, Mirna Vázquez-Rosas-Landa, Gabriel Yaxal Ponce-Soto, Alejandra Moreno-Letelier, Gabriela Olmedo-Álvarez, Luis E. Eguiarte, and Valeria Souza

Abstract In this chapter, we describe some genomic studies that have been conducted in the CCB to describe the microdiversity of different microbial taxa. Firstly, we describe the genomic variation in the endemic species *B. coahuilensis*, for which we analyzed three strains, and the evolutionary mechanism that have shaped their genomes, in particular for their adaptations to a low phosphorus environment. Secondly, we describe population genomics of *Proteobacteria* such as *Pseudomonas*, *Vibrio*, *Photobacterium*, and *Aeromonas*, finding many genes related to bacterial interactions, such as type III and VI secretion systems, which led us to propose that these systems play an important role in interactions among bacterial communities. Many of these interaction genes have been classified previously as virulence factors. However, at CCB these genes were not found in a cassette configuration—such as an integron or a pathogenicity island—in any strain suggesting that they are very ancestral adaptive responses to keep both the cohesion within the community and the cheaters out of the community. Interestingly, even if *B. coahuilensis* presents many mobile elements, these elements are rare in the surveyed *Proteobacteria*, whose genomes present many CRISPS elements. The presence of CRISPR regions may also explain why there are few ISs or plasmids since it has been reported that CRISPR spacers can act as barriers against HGT.

Z. Gomez-Lunar · G. Olmedo-Álvarez
Laboratorio de Biología Molecular y Ecología Microbiana, Departamento de Ingeniería Genética, Unidad Irapuato, Centro de Investigación y de Estudios Avanzados del IPN, Irapuato, Mexico

M. Vázquez-Rosas-Landa · G. Y. Ponce-Soto · L. E. Eguiarte · V. Souza (✉)
Laboratorio de Evolución Molecular y Experimental, Departamento de Ecología Evolutiva, Instituto de Ecología, Universidad Nacional Autónoma de México, Mexico City, Mexico
e-mail: souza@unam.mx

A. Moreno-Letelier
Jardín Botánico, Instituto de Biología Universidad Nacional Autónoma de México, Mexico City, Mexico

V. Souza et al. (eds.), *Cuatro Ciénegas Ecology, Natural History and Microbiology*, Cuatro Ciénegas Basin: An Endangered Hyperdiverse Oasis,
https://doi.org/10.1007/978-3-319-93423-5_5

Introduction

The description of microbial diversity in the Cuatro Ciénegas Basin (CCB) on the basis of 16S rDNA sequences and metagenomic sequencing has been an excellent way to start the study of the hyper-diversity and of the speciation and diversification process that shaped the microbial communities in this ancient oligotrophic environment (Souza et al. 2006; Escalante et al. 2008; Breitbart et al. 2009; Cerritos et al. 2011; Arocha-Garza et al. 2017). Besides nucleotide polymorphisms, bacterial diversity is also associated with gain of genetic material and loss events that contribute to the variation in genome size, gene content, and order (Koonin and Wolf 2008). Horizontal gene transfer (HGT) is a mechanism that increases genetic variation and may result in the addition of new genes or genetic redundancy (Mira et al. 2002; Abby and Daubin 2007). Gene duplication also increases genomic size (Mira et al. 2001).

Genomic arrangements and nucleotide polymorphisms create high genomic diversity among and within species (Ochman and Davalos 2006; Abby and Daubin 2007). The genetic variation among microorganisms at fine taxonomic scale has been studied for over two decades and is often referred to as microdiversity (Fuhrman and Campbell 1998; Moore et al. 1998; García-Martinez and Rodriguez-Valera 2000; Schloter et al. 2000; Jaspers and Overmann 2004; Brown and Fuhrman 2005; Choudhary and Johri 2011; Zimmerman et al. 2013). Microdiversity is often associated to niche differences between closely related groups of microorganism, sometimes named sub-taxa (Larkin and Martiny 2017).

Next-generation sequencing (NGS) has been helpful to identify and analyze bacterial microdiversity by sequencing and comparing complete genomes, resulting in discoveries of genes or metabolic pathways associated with environmental adaptations. Bacterial genomic studies in many cases look at strain-specific genes that may contribute to individual adaptation in specific environmental conditions (Math et al. 2012; Penn and Jensen 2012; Tian et al. 2012). By combining genomics and physiological laboratory experiments with environmental observations of microdiverse sub-taxa, we can potentially improve our ability to identify both the fundamental and realized niches of microbial taxa (Larkin and Martiny 2017).

In this chapter, we describe some genomic studies that have been conducted in the CCB to describe the microdiversity of different microbial taxa. Firstly, we describe the genomic variation in the endemic species *B. coahuilensis*, for which we analyzed three strains, and the evolutionary mechanism that have shaped their genomes. We specifically searched for adaptive traits in *B. coahuilensis* that enabled survival in a low phosphorus environment. Secondly, we describe population genomics of some of the most abundant microbial taxa in the CCB, such as *Pseudomonas*, *Vibrio*, *Photobacterium*, and *Aeromonas*. We identified genes related to bacterial interactions, such as type III and VI secretion systems, which led us to propose that these systems play an important role in interactions among bacterial communities.

Comparative Genomics of *B. coahuilensis*: An Ancient Lineage in Cuatro Ciénegas

The first genome sequenced from this ancient valley was *B. coahuilensis* strain m4-4, an endemic lineage of the CCB (Alcaraz et al. 2008). Genomic approaches allowed us to recognize some remarkable adaptations to the oligotrophic environment of the CCB (see Chap. 8).

After several years of sampling, we have collected thousands of *Bacillus* species isolates, but only three strains of the 2006 survey corresponded to *B. coahuilensis*, while hundreds of strains were from other *Bacillus* lineages (i.e., *Bacillus cereus sensu lato*, *Bacillus subtilis/pumilus*, *Bacillus amyloliquefaciens* like). These findings suggested that only a relatively small population of B. coahuilensis was present in the CCB in the Laguna Grande of the Churince system, and when that lagoon dried-out because of the overexploitation of the aquifer, the species became apparently extinct.

Bacillus coahuilensis has a small genome (ca. 3.38 Mpb), and its populations are adapted to low phosphorus availability conditions. Given these characteristics, it was important to describe its microdiversity and also identify its functional traits related to phosphorus metabolism. Comparative genomics based on the average nucleotide index (ANI) indicated high conservancy among *B. coahuilensis* genomes (98.8 and higher) (Gómez-Lunar et al. 2016). As all *B. coahuilensis* strains were isolated from an oligotrophic environment, these high ANI values were consistent with other organisms that were reported to be ecologically constrained (Konstantinidis et al. 2006). Minor differences were found in the GC content percentage. Furthermore, small differences in the number of coding DNA sequences (CDS) were observed among genomes. Their pan-genome description suggested high gene content conservation, and only 6% of genes were included in the so-called "mobilome" (Fig. 5.1b), that refers to all the genes related to mobile elements (Gómez-Lunar et al. 2016). This high genome conservation is similar to what has been observed between strains from the same species (Deng et al. 2010; Peña et al. 2010; Mann et al. 2013; Patiño-Navarrete et al. 2013).

Comparative genomics indicated that a small genome size is consistent in all three strains (Fig. 5.1a) that present around 3500 genes, 86% of them shared between the three strains. We hypothesized that a small genome size in *B. coahuilensis* reflects a genome size reduction ancestral to the group. At the beginning, we had the hypothesis that all genomes would be small at CCB, as trimmed genomes could increase bacterial fitness in a low phosphorus ecosystem due to a reduced phosphorus demand for nucleotide biosynthesis (Alcaraz et al. 2008). Additionally, recent genome sequencing indicated other *Bacillus* species are constraint in the lineage size range as well, so it is possible that *Bacillus* aquatic lineages in general exhibit smaller genomes (Hernández-González and Olmedo-Álvarez 2016a, b; Zarza et al. 2017). Also, the alignment of genomes indicated a high level of synteny (Gómez-Lunar et al. 2016). These findings suggest that evolutionary forces have been maintaining a small and highly conserved genome in *B. coahuilensis* populations of the CCB, despite the rare but relevant acquisition of genetic material.

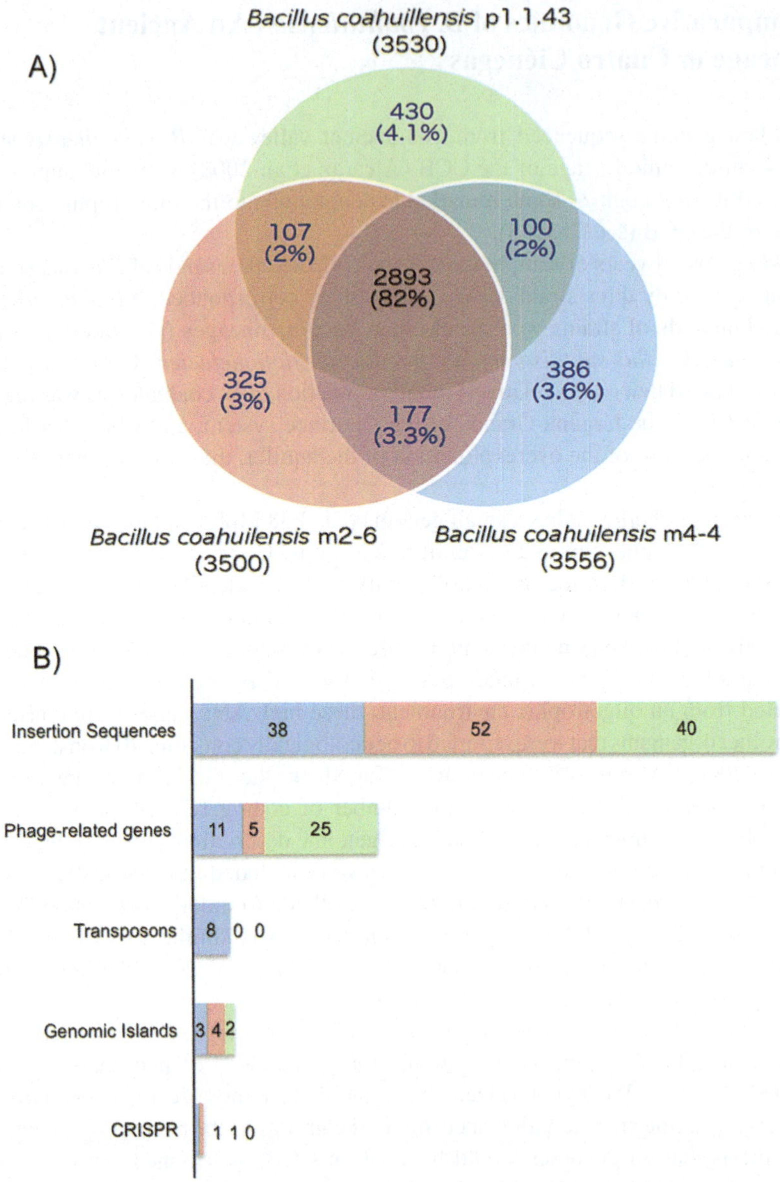

Fig. 5.1 Pan-genome of *Bacillus coahuilensis.* (**a**) Core genome and accessory genome of *B. coahuilensis.* Orthologous gene identification was performed using the bidirectional best hit (BBH) approach with mutual coverage or a shared length of 60% and an e-value $\geq 10^{-6}$. Core genome number is in black font. Accessory gene numbers are in blue font. Predicted percentages of genes are in parentheses. (**b**) Mobile genetic elements (MGEs) in *B. coahuilensis* genomes. Number in each box indicates genes related to MGEs in each genome

Reconstruction of metabolic pathways and determination of amino acid biosynthetic capabilities revealed that all *B. coahuilensis* had numerous auxotrophies to amino acids (i.e., they lack the ability to synthetize these amino acids). In all three strains, arginine auxotrophy was congruent with a predicted incomplete urea cycle. Valine, leucine, and isoleucine requirements were also consistent with an extensively eroded pyruvate pathway. Experimental determination of auxotrophies indicated a variation between strains to synthesize proline, cysteine, aspartic acid, threonine, and lysine. Glutamic acid and phenylalanine were required by all three strains. Despite observing these phenotypes, our metabolic models did not find a defective pathway to explain the auxotrophies for these seven amino acids (Gómez-Lunar et al. 2016). This result suggested that there were strain-specific regulatory differences in the expression of genes involved in the biosynthesis of these amino acids. An optional methionine biosynthesis pathway to recycle methionine was also present.

Reduction of numerous capabilities related to amino acid biosynthesis is indicative of a high dependence on nutrients available in the sediment microbial community. Absence of some biosynthetic pathways in all three genomes may be due to genome reduction and/or an adaptation to a unique niche (Alcaraz et al. 2008). This situation is called streamlining and refers to gene loss occurring in bacteria (Giovannoni et al. 2005). It has been suggested that gene loss occurs particularly in microbial communities, due to changes in the availability of public goods (Morris et al. 2012).

Carbon source utilization is also a relevant functional trait that is worth to evaluate for microorganisms isolated from low-nutrient environments. Metabolic models for carbon source utilization suggest subtle differences among *B. coahuilensis* strains, with *B. coahuilensis* p1.1.43 showing more versatility in its use of carbon sources. All *B. coahuilensis* strains studied were able to use poly- and oligosaccharides of D-glucose, D-fructose, and D-mannose, mostly linked by alpha bonds. Metabolic models suggested the presence of genes encoding alpha- and beta-glucosidases for the hydrolysis of several di- and oligosaccharides of D-glucose and D-fructose, but did not detect genes encoding galactosidase. Additionally, Biolog results supported the finding that *B. coahuilensis* strains were not able to use oligosaccharides of D-galactose. The metabolic models and Biolog results indicated that all three strains use D-ribose, D-fructose, and alpha-D-glucose. *B. coahuilensis* m4-4 was able to use more monosaccharides in comparison to *B. coahuilensis* m2-6 and *B. coahuilensis* p1.1.43.

Most metabolic models agreed with the carbohydrate utilization results obtained through Biolog. However, we found that some pathways (L-arabinose and D-xylose) could not be fully elucidated. Our findings suggest that unlike *Escherichia coli*, the standard study model in microbiology, *B. coahuilensis* may use alternative enzymes or metabolic pathways not yet described. Similar observations have been done in other non-model bacteria, such as marine bacteria *Roseobacter litoralis* (Kalhoefer et al. 2011). Utilization of alternative carbon sources, such as carboxylic acids and aromatic compounds in *B. coahuilensis*, suggest that there has been a diversification in carbohydrate scavenging capabilities among strains.

Most strain-specific genes in *B. coahuilensis* are hypothetical; however, 5–8% of them encoded mobile genetic elements (MGEs, Fig. 5.1b), most being phage-related genes, clustered regularly interspersed short palindromic repeats (CRISPR), transposons, and insertion sequences (ISs). Surprisingly, *B. coahuilensis* is one of the species in this genus with the largest number of ISs. MGE observations coincide with other bacterial lineage studies, which found that the genomes are mostly shaped by the presence of these genetic elements (Peña et al. 2010; Kalhoefer et al. 2011; Meyer and Huber 2014). Only *B. coahuilensis* m4-4 and m2-6 had CRISPR. *B. coahuilensis* m2-6 had more ORFs related to ISs than the other two genomes. More phage-related genes were present in the *B. coahuilensis* p1.1.43 genome than in those from *B. coahuilensis* m2-6 and *B. coahuilensis* m4-4 (Fig. 5.1b) (Gómez-Lunar et al. 2016). Several MGEs are located in genomic islands (GIs).

Interestingly, a GI called GIm2d had coding genes for biosynthesis of saccharide 2-keto-3-deoxymannooctulosonic acid (Kdo), an important component of lipopolysaccharide (LPS) in the cell wall of Gram-negative bacteria. Phylogenetic analysis of this region indicated that these genes are most closely related to proteins from *Gammaproteobacteria*. This phylogenetic association indicates that genes for Kdo biosynthesis were acquired by HGT. The acquisition of novel functional traits via HGT increases the potential for specific adaptations and microbial speciation in the CCB. This is a remarkable example of the microevolutionary changes in this species, and it is also an unexpected finding for Gram-positive *Firmicutes* (Gómez-Lunar et al. 2016).

Since *B. coahuilensis* lives in a low phosphorus environment, searching for functional traits related to phosphorus transport, metabolism, storage, and regulation are important to define its niche characteristics. Genomic analysis showed that all *B. coahuilensis* strains have the ability to recycle phosphorus from DNA. This ability is consistent with the presence of genes coding for enzymes regulated by phosphorus starvation involved in phosphorus recycling from extracellular DNA and RNA (Vershinina and Znamenskaya 2002; Singh et al. 2015). All three genomes encoded glycerolphosphodiesterase (GlpQ), which is important in recycling phosphorus by hydrolyzing phospholipids to glycerol-3-phosphate (Gómez-Lunar et al. 2016). In *B. subtilis,* this phosphodiesterase is also phosphorus starvation-induced and regulated by the PhoP-PhoR system (Antelmann et al. 2000). Conversely, the gene expression of *phoA* encoding alkaline phosphatase was different between strains. Alkaline phosphatase activity is constitutive for *B. coahuilensis* strain m4-4 (expression at high and low KH_2PO_4 concentrations). Alkaline phosphatase activity in *B. coahuilensis* strain p1.1.43 is only induced at low phosphate concentrations, while *B. coahuilensis* strain m2-6 does not present this enzymatic activity. Genomic analysis confirms that *B. coahuilensis* strain m2-6 does not have a gene encoding for alkaline phosphatase (Table 5.1) (Gómez-Lunar et al. 2016). These findings suggested that microdiversity also affected phosphorus utilization strategies through gene loss (i.e., alkaline phosphatase gene loss in *B. coahuilensis* strain m2-6) and changes in gene expression (gene regulation in *B. coahuilensis* strain p1.1.43).

Table 5.1 Strategies for the recycling, scavenging, and storage of phosphorus in three strains of *Bacillus coahuilensis*

Strategies	Genes	m4-4	m2-6	p1.1.43
Recycling	Alkaline phosphodiesterase I	+	+	+
	2′, 3′ cyclic nucleotide transferase (*yfkN*)	+	+	+
	Alkaline phosphatase (*phoA*)	Constitutive	–	Inducible
Scavenging	Phosphonoacetaldehyde dehydrogenase (*phnY*)	+	+	+
	Phn transporters (*phnCDE1E2*)	–	+	–
Storage or saving	Polyphosphate kinase *ppK*	+	+	+
	Exopolyphosphatase	+	+	+
	Teichoic acids biosynthesis	–	–	–
	Sulfolipid biosynthesis *sqd1*	+	+	+
	sqdX	+	+	+

Constitutive and inducible refers to *phoA* gene expression

Another strategy to access phosphorus is by scavenging alternative organic sources in an environment, wherein all *B. coahuilensis* strains have the ability to obtain phosphorus from phosphonoacetaldehyde (Table 5.1). Genomics confirm that *B. coahuilensis* genomes encoded a putative phosphonoacetaldehyde dehydrogenase (PhnY) (Table 5.1) (Gómez-Lunar et al. 2016). Unexpectedly, a phosphonate (*phn*) transport cluster was observed in *B. coahuilensis* m2-6 suggesting it has the ability to scavenge phosphorus from phosphonate (Table 5.1) (Gómez-Lunar et al. 2016). The analysis of *phn* cluster in *B. coahuilensis* m2-6 and *Bacillus* sp. m3-13 (also isolated from the CCB) revealed the presence of gene *yfkN*. Studies done with *B. subtilis* have shown that gene *yfkN* is induced in response to phosphorus starvation. The enzyme is involved in recycling phosphorus from extracellular DNA and RNA (Allenby et al. 2005). However, a study revealed *yfkN* from *Rhizobium leguminosarum* could be involved in phosphonate hydrolysis (Jonas et al. 2008). The observation of this dual function suggested that this gene could have phosphonatase activity in *B. coahuilensis* m2-6. Phylogenetic analysis suggested that the phosphonate (*phn*) transport cluster was acquired via HGT. Phosphonates have an increasingly acknowledged role in phosphorus cycling in nature. Biosynthetic and metabolic pathways that include phosphonates have been described (Quinn et al. 2007). The phosphonates occurring as a phosphorus reservoir in the ocean have been well established, as well as the strategies that marine microorganisms employ for their utilization (Villarreal-Chiu et al. 2012).

Additionally, we discovered through genomic analysis that sulfolipid biosynthesis is a conserved functional trait expressed constitutively in all *B. coahuilensis* strains. Sulfolipid biosynthesis is associated to efficiently use and store of phosphorus (Table 5.1) (see Chap. 8) (Gómez-Lunar et al. 2016). Another strategy found in all three genomes to store phosphorus is the presence of a polyphosphate kinase *ppK* gene for polyphosphate biosynthesis (Table 5.1) (Gómez-Lunar et al. 2016). For instance, a recent study of Sargasso Sea plankton indicated that an enrichment in polyphosphate coincided with an increase in alkaline phosphatase activity and a substitution of phospholipids for sulfolipids, both of which are indicators of phosphorus stress (Martin et al. 2014). Finally, the absence of genes related to teichoic acid biosynthesis is another consistent strategy to store phosphorus in *B. coahuilensis* (Table 5.1) (see Chap. 8) (Gómez-Lunar et al. 2016). Teichoic acids in Gram-positive bacteria are considered "essential" cell wall components as well as phosphorus storage molecules (Kulakovskaya 2014).

Recycling phosphorus from DNA, scavenging from phosphonates, phosphate storage, and the ability to synthesize membrane sulfolipids are strategies that help *B. coahuilensis* survive in an environment extremely limited in phosphorus. The evolutionary force explaining the diversification or niche differentiation in phosphorus acquisition strategies could be fine local adaptation or just genetic drift. Sometimes restricted dispersal creates geographically structured subpopulations (each with a reduced N_e), promoting their genetic diversification (Luo and Moran 2015).

Only three strains of *B. coahuilensis* have been recovered in the last 12 years of extensive isolation efforts, suggesting that this species might be structured in a subpopulation with a restricted dispersal that could promote microevolutionary changes and some niche differentiation. Our findings show that the microdiversity of *B. coahuilensis* is the result not only from gene loss and/or gain but also from gene regulation, affecting specific pathways. *B. coahuilensis* lineages have acquired different strategies, such as recycling, scavenging, and storage, and seem to occupy different ecological niches in order to survive in an environment with a remarkable low phosphorus concentration.

Population Genomics of *Gammaproteobacteria* from Cuatro Ciénegas

The *Gammaproteobacteria* class comprises some of the most studied bacterial groups at genetic and genomic levels. The class includes *Escherichia coli*, an organism that has been the study model in prokaryotes. This class also includes genera *Vibrio*, *Aeromonas*, and *Pseudomonas* that have been also well studied as they are of great relevance from a clinical and economic perspective. In particular, *Vibrio cholerae* is known worldwide for causing the infectious disease cholera that is estimated to affect around 1.3 to 4.0 million people each year (Ali et al. 2015). *V. cholerae* is also associated with the early mortality syndrome or acute hepatopancreatic disease in shrimp resulting in the loss of millions of dollars in the aquaculture industry (Choi et al. 2017). The genus *Aeromonas* was previously a member of the *Vibrionaceae* family. Colwell et al. (1986) proposed reclassifying it in a new family called *Aeromonadaceae* due to important differences, such as its high GC content. On the other hand, *Pseudomonas* species are ubiquitous, as they can be found in soil, aquatic environments, and plant or animal tissues. *Pseudomonas* spp. are known for their metabolic versatility and genetic plasticity. *Pseudomonas aeruginosa* is a major opportunistic human pathogen and causes bacteremia in immunocompromised patients (Stover et al. 2000). The natural resistance of *P. aeruginosa* to antibiotics makes its infections challenging to eradicate. Some *Pseudomonas* strains have also been reported to produce metabolites that stimulate plant growth or inhibit plant pests (de Souza et al. 2003; Yoshida et al. 2012). Some products of *Pseudomonas* metabolism are of industrial interest, such as biosurfactants (Chong and Li 2017; Henkel et al. 2017).

The great diversity within *Gammaproteobacteria* makes this group of particular interest not only from a clinical or economic point of view but also from an evolutionary perspective. *Gammaproteobacteria* has been consistently isolated from the CCB (Escalante et al. 2008, 2009; Rodriguez-Verdugo et al. 2012; Ponce-Soto et al. 2015; Vazquez-Rosas-Landa et al. 2017). We decided to further study this class and in particular to sequence members from *Vibrio* and its related groups *Photobacterium*, *Aeromonas*, and *Pseudomonas*. We intended to gain insights into their evolution via structure and genomic content analysis.

Bacteria have a broad range of strategies to interact with their environment as well as with their community. However, many of the genes associated with these interactions have been defined as pathogenic elements. As previously mentioned, *Vibrio* and *Pseudomonas* are widely known for their clinical relevance with *Vibrio* being an infectious agent, and *Pseudomonas* being an opportunistic pathogen. Both genera share features that allow them to infect their hosts. While studying the ecological and evolutionary forces involved in the generation and maintenance of diversity in Cuatro Ciénegas, we decided to also analyze the so-called pathogenic elements present in their genomes from an ecological perspective (Vazquez-Rosas-Landa et al. 2017).

We analyzed the presence of virulence factors in 51 sequenced strains from the CCB and 8 reference strains, in order to define a so-called virulence core (Vazquez-Rosas-Landa et al. 2017). Within the different genera analyzed in this study, we found a broad range of virulence factors in the CCB strains. The conservation of these virulence-related genes in free-living bacteria from a relatively pristine and isolated environment suggests that they are not inherently related to pathogenicity. Instead, these genes seem to be part of the adaptations of free-living microbes related to their particular ecological niches used to interact with the environment and other bacteria, including genes related to chemotaxis and flagella.

When the 16S rDNA tree was compared to the "virulent core" phylogeny, a shift in the phylogenetic position of three *Vibrio* strains was detected, indicated HGT. Individual phylogenies of core genes showed a second copy of the flagella gene *flhA* in *Vibrio parahaemolyticus*. This gene could represent either another HGT event or, more likely, a duplication that occurred before the divergence of these groups based on sequence grouping. In relation to the type III secretion system (T3SS) core elements, the seven evolutionary lineages of this system are conserved, and the CCB strains do not constitute a separate group, and another probable HGT event is proposed among *Vibrio* clade II and T3SS2 of *V. parahaemolyticus* as well as with one *Aeromonas* strain. In contrast to *Aeromonas* and *Vibrio*, few T3SS components were found in *Pseudomonas* associated with the CCB. The findings suggest that purifying selection has eroded the type III secretion system in most of the CCB strains, or that there are undescribed homologs.

The virulence factors among the analyzed CCB strains show individual genes associated with virulence, such as toxin and antibiotic production and resistance. However, these different virulence genes were not found in a cassette configuration—such as an integron or a pathogenicity island—in any strain (Vazquez-Rosas-Landa et al. 2017). This study showed us that despite the pathogenic classification of some genes, this may not always be the case. These genes are likely playing an ecological role in its natural environment, either in signaling or motility, and are critical to define their niches, in particular their Eltonian component.

Microorganisms carrying antibiotic resistance genes tend to be more abundant in human-impacted environments. Some researchers consider that this is the result of a human-mediated dispersal of resistance genes and resistant bacteria (see, for instance, Cristobal-Azkarate et al. 2014). Antibiotic resistance is commonly associated with mobile genetic elements (MGEs), such as genomic islands, integrons, or bacteriophages (Dobrindt et al. 2004; Hazen et al. 2010). Integrons are gene-capturing elements that can host a wide range of "cassettes" that promote the spread of antibiotic resistance genes or virulence factors (Cury et al. 2016). However, the ecological and evolutionary roles of MGEs are complex. For example, not all genomic islands are pathogenicity islands, as well as most plasmids are not related to pathogenesis (Dobrindt et al. 2004). Integrons and the genes associated may harbor virulent genes, as well as resistance genes (Mazel et al. 1998).

As is well known, phages are not only one of the most efficient mechanisms of HGT (Jiang and Paul 1998; Canchaya et al. 2003; Weinbauer and Rassoulzadegan 2004; Moon et al. 2016), but they are also their most efficient pathogens of bacteria, controlling in many cases their population dynamics, ecology, and evolution (Weitz and Wilhelm 2012). CRISPR loci are tightly associated with phage resistance (Bland et al. 2007). Isolates from the *Proteobacteria* from CCB contained few mobile elements besides CRISPR sequences that prevent virus from transferring genes (Fig. 5.2) since few *Vibrio* presented integrons and phages (Vazquez-Rosas-Landa et al. 2017). The presence of CRISPR regions may also explain why there are few ISs or plasmids since it has been reported that CRISPR spacers can act as barriers against HGT (García-Gutiérrez et al. 2015; Sheludchenko et al. 2015). Our findings of a strong cohesion of *Vibrio* lineages within CCB given the reproductive isolation between lineages (Vazquez-Rosas-Landa et al. 2017) support the idea that the costs for extra DNA replication could be a counterselection force against foreign DNA incorporation in this oligotrophic environment (Souza et al. 2012).

Fig. 5.2 Schematic representation of the different mobile genetic elements in *Gammaproteobacteria* from Cuatro Ciénegas. 16S rDNA phylogenetic reconstruction. Inner circles (from the inside outward) represent the presence of integrons, phages, and CRISPR elements in the analyzed genomes

Conclusions and Perspectives for the Genomic Studies of Bacteria in Cuatro Ciénegas

Genomic analysis of *Bacillus* and of the different *Gammaproteobacteria* and the physiological and genetic experiments in *Bacillus*, along with environmental observations, have shown important adaptations to life in an oligotrophic environments, in particular on how to deal with the scarcity of phosphorus and how to start analyzing the fundamental and realized niches and their Eltonian and Grinnellian (see Soberón, 2007 and Chap. 6 in this book) components in microbial taxa. So far, we know that there is a disproportion of the metagenomes associated, not to P acquisition, but to microbial interactions suggesting that the Eltonian niche is profoundly important in this site. In the near future, population genomic studies (Vazquez-Rosas-Landa et al. 2017) will allow us to better disentangle and analyze the fine details of niche micro-differences as well as local adaptations among strains. For instance, in her Ph.D. thesis, Vazquez-Rosas-Landa (2017) compared the genomes of 42 *Vibrionaceae* representing 6 different lineages of the CCB *Vibrionaceae*, most of them specialist of either water or sediment. She found that while all the specialist lineages presented a closed pan-genome and recombination restricted only to their closely related strains, the generalistic clade had an open pan-genome as well as an internal structure with at least three sub-clades, suggesting that it is in the process of diversification. These features are an indication that the strains have a broader habitat spectrum explaining their success in this harsh environment. The same cannot be said of the extreme specialist *B. coahuilensis*, which got extinct along with its particular community that belonged to the desiccation lagoon that no longer exists in the Churince.

Acknowledgments The research was funded in part by Papiit-UNAM project IG200215 for *Vibrio* comparative genomics to VS and LEE, and CONACYT CB-2013-01 No. 220536 to GO-A. This chapter was written during a sabbatical leave of VS in the Department of Ecology, Evolution, and Behavior, University of Minnesota in Dr. Michael Travisano´s laboratory, and of LEE, in the Department of Plant and Microbial Biology, University of Minnesota in Dr. Peter Tiffin's laboratory, both supported by the program PASPA-DGAPA, UNAM.

References

Abby S, Daubin V (2007) Comparative genomics and the evolution of prokaryotes. Trends Microbiol 15(3):135–141

Alcaraz LD et al (2008) The genome of *Bacillus coahuilensis* reveals adaptations essential for survival in the relic of an ancient marine environment. Proc Natl Acad Sci 105(15):5803–5808

Ali M, Nelson AR, Lopez AL, Sack D (2015) Updated global burden of cholera in endemic countries. PLoS Negl Trop Dis 9(6):e0003832

Allenby NEE et al (2005) Genome-wide transcriptional analysis of the phosphate starvation stimulon of *Bacillus subtilis*. J Bacteriol 187(23):8063–8080

Antelmann H, Scharf C, Hecker M (2000) Phosphate starvation-inducible proteins of *Bacillus subtilis*: proteomics and transcriptional analysis. J Bacteriol 182(16):4478–4490

Arocha-Garza HF et al (2017) High diversity and suggested endemicity of culturable Actinobacteria in an extremely oligotrophic desert oasis. PeerJ 5:e3247

Bland C et al. (2007) CRISPR Recognition Tool (CRT): a tool for automatic detection of clustered regularly interspaced palindromic repeats. BMC Bioinformatics 8(1):209(1–19) https://doi.org/10.1186/1471-2105-8-209

Breitbart M et al (2009) Metagenomic and stable isotopic analyses of modern freshwater microbialites in Cuatro Ciénegas, Mexico. Environ Microbiol 11(1):16–34

Brown M, Fuhrman J (2005) Marine bacterial microdiversity as revealed by internal transcribed spacer analysis. Aquat Microb Ecol 41(1):15–23

Canchaya C, Fournous G, Chibani-Chennoufi S, Dillmann ML, Brussow H (2003) Phage as agents of lateral gene transfer. Curr Opin Microbiol 6(4):417–424

Cerritos R et al (2011) Diversity of culturable thermo-resistant aquatic bacteria along an environmental gradient in Cuatro Cienegas, Coahuila, Mexico. Antonie Van Leeuwenhoek 99(2):303–318

Choi M, Stevens AM, Smith SA, Taylor DP, Kuhn DD (2017) Strain and dose infectivity of *Vibrio parahaemolyticus*: the causative agent of early mortality syndrome in shrimp. Aquac Res 48:3719–3727

Chong H, Li Q (2017) Microbial production of rhamnolipids: opportunities, challenges and strategies. Microb Cell Factories 16:137

Choudhary DK, Johri BN (2011) Ecological significance of microdiversity: coexistence among casing soil bacterial strains through allocation of nutritional resource. Indian J Microbiol 51(1):8–13

Colwell RR, MacDonell MT, DeLey J (1986) Proposal to recognize the family Aeromonadaceae fam. nov. Int J Syst Bacteriol 36:473–477

Cristobal-Azkarate J, Dunn JC, Day JMW, Amabile-Cuevas CF (2014) Resistance to antibiotics of clinical relevance in the fecal microbiota of Mexican wildlife. PLoS ONE 9(9):e107719. https://doi.org/10.1371/journal.pone.0107719

Cury J et al. (2016) Identification and analysis of integrons and cassette arrays in bacterial genomes. Nucleic Acids Research 4(10):4539–4550.

de Souza JT, De Boer M, De Waard P, Van Beek TA, Raaijmakers JM (2003) Biochemical, genetic, and zoosporicidal properties of cyclic lipopeptide surfactants produced by *Pseudomonas fluorescens*. Appl Environ Microbiol 69:7161–7172

Deng X et al (2010) Probing the pan-genome of *Listeria monocytogenes*: new insights into intraspecific niche expansion and genomic diversification. BMC Genomics 11:500

Dobrindt U, Hochhut B, Hentschel U, Hacker J (2004) Genomic islands in pathogenic and environmental microorganisms. Nat Rev Microbiol 2(5):414–424. https://doi.org/10.1038/nrmicro884

Escalante AE et al (2008) Diversity of aquatic prokaryotic communities in the Cuatro Cienegas basin. FEMS Microbiol Ecol 65(1):50–60

Escalante AE et al (2009) *Pseudomonas cuatrocienegasensis* sp. nov., isolated from an evaporating lagoon in the Cuatro Cienegas valley in Coahuila, Mexico. International Journal of Systematic and Evolutionary Microbiology 59(6):1416–1420.

Fuhrman JA, Campbell L (1998) Microbial microdiversity. Nature 393(6684):410–411

García-Martinez J, Rodriguez-Valera F (2000) Microdiversity of uncultured marine prokaryotes: the SAR11 cluster and the marine Archaea of Group I. Mol Ecol 9(7):935–948

García-Gutiérrez E et al. (2015) CRISPR Content Correlates with the Pathogenic Potential of Escherichia coli. PLoS ONE 10(7):e0131935. https://doi.org/10.1371/journal.pone.0131935

Giovannoni SJ et al (2005) Genome streamlining in a cosmopolitan oceanic bacterium. Science 309(5738):1242–1245

Gómez-Lunar Z et al (2016) Microevolution analysis of *Bacillus coahuilensis* unveils differences in phosphorus acquisition strategies and their regulation. Front Microbiol 7:58

Hazen TH, Pan L, Gu JD, Sobecky PA (2010) The contribution of mobile genetic elements to the evolution and ecology of *Vibrios*. FEMS Microbiol Ecol 74(3):485–499. https://doi.org/10.1111/j.1574-69412010.00937.x

Henkel M, Geissler M, Weggenmann F, Hausmann R (2017) Production of microbial biosurfactants: status quo of rhamnolipid and surfactin towards large-scale production. Biotechnol J 12(7):28544628. https://doi.org/10.1002/biot.201600561

Hernández-González IL, Olmedo-Álvarez G (2016a) Draft whole-genome sequence of the type strain *Bacillus horikoshii* DSM 8719. Genome Announc 4(4):e00640–e00616

Hernández-González IL, Olmedo-Álvarez G (2016b) Draft whole-genome sequence of the type strain *Bacillus aquimaris* TF12 T. Genome Announc 4(4):e00641–e00616

Jaspers E, Overmann J (2004) Ecological significance of microdiversity: identical 16S rRNA gene sequences can be found in bacteria with highly divergent genomes and ecophysiologies. Appl Environ Microbiol 70(8):4831–4839

Jiang SC, Paul JH (1998) Gene transfer by transduction in the marine environment. Appl Environ Microbiol 64(8):2780–2787

Jonas S et al (2008) A new member of the alkaline phosphatase superfamily with a formylglycine nucleophile: structural and kinetic characterisation of a phosphonate monoester hydrolase/phosphodiesterase from *Rhizobium leguminosarum*. J Mol Biol 384(1):120–136

Kalhoefer D et al (2011) Comparative genome analysis and genome-guided physiological analysis of *Roseobacter litoralis*. BMC Genomics 12:324

Konstantinidis KT, Ramette A, Tiedje JM (2006) The bacterial species definition in the genomic era. Philos Trans R Soc B 361(1475):1929–1940

Koonin EV, Wolf YI (2008) Genomics of bacteria and archaea: the emerging dynamic view of the prokaryotic world. Nucleic Acids Res 36(21):6688–6719

Kulakovskaya T (2014) Phosphorus storage in microorganisms: diversity and evolutionary insight. Biochem Physiol 4(1):e130

Larkin AA, Martiny AC (2017) Microdiversity shapes the traits, niche space, and biogeography of microbial taxa. Environ Microbiol Rep 9(2):55–70

Luo H, Moran MA (2015) How do divergent ecological strategies emerge among marine bacterioplankton lineages? Trends Microbiol 23(9):577–584

Mann RA et al (2013) Comparative genomics of 12 strains of *Erwinia amylovora* identifies a pan-genome with a large conserved core. PLoS One 8(2):e55644

Martin P et al (2014) Accumulation and enhanced cycling of polyphosphate by Sargasso Sea plankton in response to low phosphorus. Proc Natl Acad Sci 111(22):8089–8094

Math RK et al (2012) Comparative genomics reveals adaptation by *Alteromonas* sp. SN2 to marine tidal-flat conditions: cold tolerance and aromatic hydrocarbon metabolism. PLoS One 7(4):e35784

Mazel D, Dychinco B, Webb VA, Davies J (1998) A distinctive class of integron in the *Vibrio cholerae* genome. Science 280(5363):605–608

Meyer JL, Huber JA (2014) Strain-level genomic variation in natural populations of *Lebetimonas* from an erupting deep-sea volcano. Int Soc Microb Ecol J 8(4):867–880

Mira A, Klasson L, Andersson SGE (2002) Microbial genome evolution: sources of variability. Curr Opin Microbiol 5(5):506–512

Mira A, Ochman H, Moran NA (2001) Deletional bias and the evolution of bacterial genomes. Trends Genet 17(10):589–596

Moon BY, Park JY, Robinson DA, Thomas JC, Park YH, Thornton JA, Seo KS (2016) Mobilization of genomic islands of *Staphylococcus aureus* by temperate bacteriophage. PLoS ONE 11(3):e0151409. https://doi.org/10.1371/journal.pone.0151409

Moore LR, Rocap G, Chisholm SW (1998) Physiology and molecular phylogeny of coexisting *Prochlorococcus* ecotypes. Nature 393(6684):464–467

Morris JJ, Lenski RE, Zinser ER (2012) The black queen hypothesis: evolution of dependencies through adaptive gene loss. MBio 3(2):e00036-12

Ochman H, Davalos LM (2006) The nature and dynamics of bacterial genomes. Science 311(5768):1730–1733

Patiño-Navarrete R et al (2013) Comparative genomics of *Blattabacterium cuenoti*: the frozen legacy of an ancient endosymbiont genome. Genome Biol Evol 5(2):351–361

Peña A et al (2010) Fine-scale evolution: genomic, phenotypic and ecological differentiation in two coexisting *Salinibacter ruber* strains. Int Soc Microb Ecol J 4(7):882–895

Penn K, Jensen PR (2012) Comparative genomics reveals evidence of marine adaptation in *Salinispora* species. BMC Genomics 13:86

Ponce-Soto GY, Aguirre-von-Wobeser E, Eguiarte LE, Elser JJ, Lee ZMP, Souza V (2015) Enrichment experiment changes microbial interactions in an ultra-oligotrophic environment. Front in Microbiology 6(246). https://doi.org/10.3389/fmicb.2015.00246

Quinn JP et al (2007) New ways to break an old bond: the bacterial carbon-phosphorus hydrolases and their role in biogeochemical phosphorus cycling. Environ Microbiol 9(10):2392–2400

Rodríguez-Verdugo A, Souza V, Eguiarte LE, Escalante AE (2012) Diversity across seasons of culturable *Pseudomonas* from a desiccation lagoon in Cuatro Cienegas, Mexico. International Journal of Microbiology 212: 10. https://doi.org/10.1155/2012/201389

Schloter M et al (2000) Ecology and evolution of bacterial microdiversity. FEMS Microbiol Rev 24(5):647–660

Sheludchenko MS, Huygens F, Stratton H, Hargreaves M (2015) CRISPR Diversity in E. coli Isolates from Australian Animals, Humans and Environmental Waters. PLoS ONE 10(5):e0124090. https://doi.org/10.1371/journal.pone.0124090

Singh DN et al (2015) Identification and characterization of a novel phosphodiesterase from the metagenome of an Indian coalbed. PLoS One 10(2):e0118075

Soberón J (2007) Grinnellian and Eltonian niches and geographic distributions of species. Ecol Lett, 10:1115–1123

Souza V et al (2006) An endangered oasis of aquatic microbial biodiversity in the Chihuahuan desert. Proc Natl Acad Sci 103(17):6565–6570

Souza V, Siefert JL, Escalante AE, Elser JJ, Eguiarte LE (2012). The Cuatro Ciénegas Basin in Coahuila, Mexico: An Astrobiological Precambrian Park. Astrobiology, 12(7):641–647. https://doi.org/10.1089/ast.2011.0675

Stover CK et al (2000) Complete genome sequence of *Pseudomonas aeruginosa* PAO1, an opportunistic pathogen. Nature 406(6799):959

Tian CF et al (2012) Comparative genomics of rhizobia nodulating soybean suggests extensive recruitment of lineage-specific genes in adaptations. Proc Natl Acad Sci 109(22):8629–8634

Vázquez-Rosas-Landa M, Yaxal Ponce-Soto G, Eguiarte LE, Souza V (2017) Comparative genomics of free-living Gammaproteobacteria: pathogenesis-related genes or interaction-related genes?. FEMS Pathogens and Disease 75(5):1–12

Vélez P, Gasca-Pineda J, Rosique-Gil E, Eguiarte LE, Espinosa-Asuar L, Souza V (2016) Microfungal oasis in an oligotrophic desert: diversity patterns and community structure in three freshwater systems of Cuatro Ciénegas, Mexico. PeerJ, 4, e2064. https://doi.org/10.7717/peerj.2064

Vershinina OA, Znamenskaya LV (2002) The pho regulons of bacteria. Microbiology 71(5):497–511

Villarreal-Chiu JF, Quinn JP, McGrath JW (2012) The genes and enzymes of phosphonate metabolism by bacteria, and their distribution in the marine environment. Front Microbiol 3:1–13

Weinbauer MG, Rassoulzadegan F (2004) Are viruses driving microbial diversification and diversity? Environ Microbiol 6(1):1–11

Weitz JS, Wilhelm SW (2012) Ocean viruses and their effects on microbial communities and biogeochemical cycles. F1000 Biol Rep 4:17. https://doi.org/10.3410/B4-17

Yoshida S, Ohba A, Liang YM, Koitabashi M, Tsushima S (2012) Specificity of *Pseudomonas* isolates on healthy and *Fusarium* head blight-infected spikelets of wheat heads. Microb Ecol 64(1):214–225. https://doi.org/10.1007/s00248-012-0009-y

Zarza E, Alcaraz LD, Aguilar-Salinas B, Islas A, Olmedo-Álvarez A (2017) Complete genome sequence of *Bacillus horikoshii* strain 20a from Cuatro Cienegas, Coahuila, Mexico. Genome Announc 5(30):e00592-17

Zimmerman AE, Martiny AC, Allison SD (2013) Microdiversity of extracellular enzyme genes among sequenced prokaryotic genomes. ISME J 7(6):1187–1199

Chapter 6
The Niche at the Edge of Life or the Microbial Ecology (Including Microfungi) of Cuatro Ciénegas: Mutualisms with Locals, Antagonisms Against Foreigners

Patricia Vélez, Laura Espinosa-Asuar, Michael Travisano, Luis E. Eguiarte, and Valeria Souza

Abstract In this chapter we will explore whether the niche concept is useful to understand microbes, since we believe that this key concept in ecology is particularly useful in the investigation of microbial biodiversity. On the one hand, with microbes we can conduct experiments to analyze in detail all the aspects of the Grinnellian niche, in a way that cannot be done in other organisms. On the other hand, even if complicated, we can develop different experiments and observations to start analyzing the Eltonian aspects of the niche. Here we develop our ideas on how to study the niche components in microbial systems and describe our advances in analyzing these components in a particular group of organisms in the Cuatro Ciénegas Basin (CCB). In this oasis, the distinctive oligotrophic conditions of several freshwater systems in CCB promote unique microbial interactions in order to cope with nutrient stress, where antagonist and synergic effects influence bacterial diversity structure, modifying the nature and intensity of biotic interactions. For example, *Bacillus* has few antagonistic interactions among strains of a given sampling site but strong antagonisms when competing against strains from different sites. Interestingly, antagonistic strains belonged to two genetically close species, and their targets belonged to taxonomically separate groups. Even more so, the same *Bacillus* strains showed great diversity in the dependence of the different

P. Vélez
Jardín Botánico, Instituto de Biología Universidad Nacional Autónoma de México, Mexico City, Mexico

L. Espinosa-Asuar · L. E. Eguiarte · V. Souza (✉)
Laboratorio de Evolución Molecular y Experimental, Departamento de Ecología Evolutiva, Instituto de Ecología, Universidad Nacional Autónoma de México, México City, Mexico
e-mail: souza@unam.mx

M. Travisano
Department of Ecology, Evolution and Behavior, University of Minnesota, Saint Paul, MN, USA

V. Souza et al. (eds.), *Cuatro Ciénegas Ecology, Natural History and Microbiology*, Cuatro Ciénegas Basin: An Endangered Hyperdiverse Oasis,
https://doi.org/10.1007/978-3-319-93423-5_6

strains on other community members. While antagonistic strains were in general prototrophs, a large proportion of sensitive and resistant strains exhibited auxotrophy suggesting that they need their community in order to survive. At CCB we believe that the cohesion of the community through these positive interactions is paramount to explain survival in such a harsh environment. As a result, the multidimensional space where the theater of life plays in CCB is very rich and full of actors, which shifts constantly, as virus and other process "chop the head" of the dominant species and former "underdogs" of the rare biosphere take their roles. Finally, since this play of "life at the edge" has been continued in the same site for a very long time with coevolving communities, here the "conversations and negotiations" between microscopic actors seem to be the reason why this stage has persisted for such a long time.

What Is a Niche? Is it a Useful Concept in Microbes?

In ecology, the niche concept is both keystone and, paradoxically, a blur. What makes the niche so elusive is that in many ways, it represents an abstract, a summary of the complete ecology of an organism. The niche was meant as a description of the complete range of conditions and resources where any given organism could live and reproduce (see review in Soberon 2007). To make things more complicated, there are at least two schools of thought on the niche concept; following Soberon (2007), we will concentrate here in the Grinnellian vs. the Eltonian concepts. Joseph Grinnell, an American ornithologist, in 1917 wrote a paper where he used the word niche as a description of the conditions, mostly abiotic (physical and chemical) where an organism can grow, survive, and reproduce. Some of these conditions can be used to predict the potential geographic distribution patterns, or potential niches, of different species of macroorganisms at large spatial scale (Soberon 2007).

Charles S. Elton, a British ecologist, working mostly with animals like mammals, wrote in 1927 a book where he considered the role of "food and enemies" in determining the fate of animal populations locally. Hence, this Eltonian niche concept is related more to the interactions among species, and with the trophic web and the footprint of the organisms into the environment (Soberon 2007). However, in the praxis, even if the Elton niche is important for a given species, there are very few works that formally analyze it, mainly because of the complexity of the biological interactions and because these are very dynamic in space and time (Comte et al. 2016). In contrast, there is a recent avalanche on Grinnellian distribution analyses; for instance, in the *Web of Science*, we found 13,457 papers on the niche subject, until February 2018, most of them related to potential niche and distribution analyses.

One important problem for the empirical study of the Eltonian niche is that its effects are local and change in time and that species interactions are harder to analyze and explain, as well as harder to measure, than to just model climate in a map. Interactions are subtle and can change depending of the context; sometimes they are accounted for as a prey-predator dynamics, but many times they can involve chemical substances that diffuse in the environment. An example of this complexity is the inhibition of growth by antibiotics, or the abundance of chemical compounds that plants and animals use, such as volatiles in a flower, employed to attract a pollinator or to evade herbivory or to attract natural enemies of the herbivores. Many more examples exist on odors, sounds, and visual clues that plants and animals emit to signal their presence and their reproductive status.

Nonetheless, the capacity of dispersal of an organism is also relevant for the niche amplitude and distribution. In the small world of microbes, to evaluate and understand the true capabilities of distribution or a given lineage can be tricky: even if some spores (or other dispersal structure) do in fact have the potential to cross the globe dragged out by wind and survive, most of the microbes do not "fly" since they can be very prone to desiccation and are sensitive to UV light. In particular, aquatic microbes can disperse with water, if the water maintains the physicochemical properties such as pH, salinity, and temperature range that are needed in its Grinnellian niche specifications.

Most bacteria or fungi that have been deemed as cosmopolitan in their distribution, after carefully modern genetic analyses, are shown to have been either moved by humans (as is the case of *E. coli*, Souza et al. 2012) or to present local genotypes, even in the water column of the ocean (Flombaum et al. 2013).

We believe that in this microbial world, species interactions are very important, more than usually acknowledge or understood. For instance, in the world of microarthropods, plankton, microalgae, fungi, and bacteria, chemical signals travel for a short distance, and local differentiation is mostly created by the own community and the local conditions that surround them at the millimeter-micron scale (de Anda 2017). These local conditions along with the local interactions constitute microniches that, as in a fractal vision of ecology, determine local survivorship (Stubbendieck 2016).

The microbiome is the key biotic compartment for nutrient cycling in the Earth, dominates all the biogeochemical cycles, and plays particular and irreplaceable roles, so it is of great importance to explore the niche concept in the small world.

To answer the question at the beginning of this chapter of whether the niche concept is useful to understand microbes, we believe that it is a particularly useful concept and prone to investigation in microorganisms. On the one hand, we can conduct experiments to analyze in detail all the aspects of the Grinnellian niche, in a way that cannot be done in other organisms. On the other hand, even if complicated, we can develop different experiments and observations to start analyzing the Eltonian aspects of the niche. Here we develop our ideas on how to study the niche components in microbial systems and describe our advances in analyzing these components in a particular group of organisms in the Cuatro Ciénegas Basin (CCB).

Why Bacteria and Other Microorganisms Are Relevant and Interesting for Niche Studies?

Bacteria and other microbes (in particular single cell fungi, as yeast) have several advantages for studying niche variables over the classic subject of niche studies: birds, mammals, fishes, reptiles, insects, or plants. Their obvious main advantage is that they are far more amenable to experimental manipulation. Not only can they attain large population sizes in a small tube, but several, even hundreds of replicates of the experiments can be easily designed. Also, microorganisms reach thousands of generations in a relatively short time, so it is possible to analyze not only what happens in a given moment but also explore how a given niche variable will behave in the future, and by doing so, we can also study its evolution.

All this is particularly exciting since current standard niche analyses can only speculate on how the distribution for some particular taxa was in the past by using ancestral state reconstructions of the niche. Ecologists, by invoking niche conservatism lose an important part of the picture. Not only are they missing the dynamics within food webs and local adaptation, as is the case of invasive species in a new environment (Comte et al. 2016) but, since the niche conservatism null model implicitly minimizes the role of evolution, end up with a static and limited picture.

Encompassing both niche models, Hutchinson (1965) wrote "*The Ecological Theater and the Evolutionary Play*," where the theater of life is its multidimensional scenario. In this metaphor, ecology would be able to see one scene, while evolution can see the whole play, as the scenario changes with time and interactions. In this metaphor, for instance, microbes in experimental evolution allow us to set up particular "stages" of initial conditions and resources, in order to simplify the "play" and deduct each part.

As mentioned above, in the laboratory we can have replicates and control and change most of the Grinnellian niche variables. We can grow replicates of the same organism for a thousand of generations in gradients of temperature, oxygenation, different gases, exposing them to changing gravity, and different types and concentrations of salts, different pH, light, and other forms of radiation. We can even have glimpses of the microorganisms Eltonian niche by manipulating the resources, such as low or high levels of carbon and other nutrients, expose them to low or high densities of their own species or put them to compete with other species in structured or unstructured environments, or include predators and other enemies (Stubbendieck 2016).

Ecological Interactions in Microbes

In all ecosystems, microorganisms usually occur in close vicinity with each other, entangling complex networks of intra- and interspecific interactions triggered by environmental variables involving elements of both Grinnellian and Eltonian niches. Indeed, microorganisms exhibit a stunning array of social behaviors (Crespi 2001; Lazdunski et al. 2004), which are commonly mediated through the production of assorted chemical compounds, generally aimed at reducing fitness of competitors (Cordero et al. 2012). However, in recent years, cooperation has been revealed as a determinant of microbial community structure as well (Gulis and Suberkropp 2003; Schuster et al. 2003; West et al. 2007). Furthermore, strong evidence suggests a genetic basis for these interactions in prokaryotic systems, as similar genotypes are more likely to cooperate, and different genotypes tend to compete (Mitri and Foster 2013).

Nutrient conditions, especially phosphorus (P) and nitrogen (N) concentrations, represent key factors when modeling microbial interactions (Gulis and Suberkropp 2003; Danger et al. 2013; Sridhar and Bärlocher 2000), particularly in oligotrophic environments (Ponce-Soto et al. 2015).

Bacterial Interactions and Niche Studies in Cuatro Ciénegas

Distinctive oligotrophic conditions of several freshwater systems in CCB promote unique microbial interactions in order to cope with nutrient stress, where antagonist and synergic effects influence bacterial diversity structure, modifying the nature and intensity of biotic interactions (Aguirre-Von-Wobeser et al. 2014).

In several experiments, we have observed that bacteria-bacteria interactions in CCB showed both antagonistic and cooperation mechanisms, in particular in different *Proteobacteria* groups (Ponce-Soto et al. 2015; Zapién-Campos et al. 2015). Among the substances that seem to regulate these interactions are cyclo-peptides, as recently described by Martínez-Carranza et al. (2017). In bacterial communities, these substances can serve both as a "membership id" to recognize the members of the community from the foreigners and also as antibiotics.

In CCB, the *Firmicutes* genus *Bacillus* has been studied in detail in different aspects. In particular, in a careful study, Pérez-Gutiérrez et al. (2013) detected few antagonistic interactions among strains of a given sampling site but strong antagonisms when competing against strains from different sites. Interestingly, antagonistic strains belonged to two genetically close species, and their targets belonged to taxonomically separate groups. Other studies using the same *Bacillus* strains showed great diversity in the dependence of the different strains on other community members. While antagonistic strains were prototrophs, a large proportion of sensitive and resistant strains exhibited auxotrophy (Rodríguez-Torres et al. 2017) suggesting that there are simultaneously antagonistic interactions and positive interactions among the members of the community to fulfill their basic nutrient requirements.

The above observations are supported for the rest of the communities by the metagenomic studies of CCB, where genes associated to interactions are commonly found in the samples. These genes are related to both antagonistic and mutualistic interactions (Peimbert et al. 2012; de Anda 2017) and seem to be mostly "balanced" under normal conditions (Moreno-Fontealba 2016). However, in the microbial mats associated with drought conditions in the Churince system, more antagonist interactions were observed in a network analysis (de Anda 2018), suggesting that the depletion of the aquifer has consequences in the structure of the community. One explanation would be that stress from drought itself induces metabolic changes (affecting nutrient status, gas exchange, etc.), including expression of more antagonistic interaction genes thus enriching for microorganisms that can respond to these conditions. Additionally, since soil compaction depends on water volume, it is possible that the spatial scale for the interactions will change with a reduced volume under draught conditions, forcing closer interactions among residents. Although the effect of drought on microbial function has not been addressed on mats, it has been documented for soil, particularly regarding agricultural practices (Ghorbani et al. 2008).

Interactions Involving Microfungi in Cuatro Ciénegas

In addition to bacteria, an interesting microfungal diversity has been reported in CCB aquatic systems, including OTUs phylogenetically affiliated to several genera within *Zygomycota*, *Basidiomycota*, and *Ascomycota* (Velez et al. 2016). Interestingly, these microfungi seem to be dominated by transient aquatic taxa, as traditionally classified resident taxa have not been recorded. This transient character may be a result of the long-term strong competition with the highly adapted microbial community and high selective pressures. Nonetheless, abundantly occurring taxa are shared among different aquatic systems, suggesting these fungi are highly adapted to spatial and temporal heterogeneity (Velez et al. 2016).

Though phylogenetically very distant, bacteria and fungi coexist in the aquatic systems of the CCB, where cross-kingdom interactions have been observed to shift from synergy to antagonism, as nutrients increase in accordance to the stress-gradient hypothesis. For instance, during in vitro co-culture assays, nutrient-dependent growth changes in bacteria (members of *Aeromonas* and *Vibrio*) and fungi (*Coprinellus micaceus*, *Cladosporium* sp., and *Aspergillus niger*, which constitute an important portion of the transient aquatic fungal communities in this area) were identified (Velez et al. 2016). Evidence from these experiments suggests that fungal-bacterial interactions in the CCB depend on the nutrient source and availability, as negative interactions were promoted in rich growth media, whereas the growth medium resembling oligotrophic conditions determined an increase of positive interactions (Velez et al. 2016).

Our results suggest that cross-kingdom synergistic interactions might favor microbial colonization and growth under low-nutrient conditions, representing an adaptive trait to oligotrophic environments. Remarkably, these patterns diverge from findings for freshwater environments, where it seems that antagonistic mechanisms prevail (Wohl and McArthur 2001; Gulis and Suberkropp 2003; Mille-Lindblom and Tranvik 2003), implying unique ecological patterns for the microbial community within oligotrophic freshwater systems in the CCB.

Fungal-fungal interactions confirmed nutrient-dependent patterns for the niches of different lineages of the microbial communities in the CCB (Velez et al. 2016), as was described for *Bacillus* (Rodríguez-Torres et al. 2017; Chap. 8 of this book), where codependence of other members of the community seems to be the rule. These microbial associations may regulate ecosystem processes, stimulating the production of extracellular enzymes and natural products (Boddy 2000; Woodward and Boddy 2008).

Nonetheless, in contrast with fungal-bacterial cross-kingdom interactions, experimental evidence suggests that the microfungal community in this area might be predominantly shaped by antagonistic mechanisms under low-nutrient conditions. These results resemble observations from abundant fungal species on epilithic biofilms of limestone walls exposed to a subtropical climate, where intra- and interspecific inhibitory effects were dominantly observed under oligotrophic conditions (Morón-Ríos et al. 2017). Decisively, microfungal interactions can influence microbial colonization and community assembly in the CCB, altering further interactions in the microbial community.

Microfungi and Mite Interactions in Cuatro Ciénegas

Experiments of coexistence were conducted at the three-level trophic microbial chain using an in vitro diet preference bioassays. For this experiment, three transient, aquatic saprotrophic microfungal taxa (co-occurring with oribatid mites on test panels) identified as *Aspergillus niger, Talaromyces* sp., and *Pleosporales* sp. were offered individually and simultaneously to aquatic mites from the Churince system, finding that oribatid mites fed preferentially on the profusely developing *A. niger*, a fungi that establish beneficial cross-feeding with the local bacteria (Velez et al. 2018). This is interesting, since both the mites and the fungi have unique divergent lineages in this site.

Virus, Niches, and Microbial Interactions in CCB

If we want to add another layer to this trophic chain, we have observed a very large diversity of virus that prey on all the microbial members of the communities in CCB (Taboada et al. 2018), representing an additional axis in the Eltonian microbial niches. The virome of CCB is particularly diverse (see Chap. 4 in this book) and mostly attuned to prey on bacteria, microarthropods, microalgae, as well as fungi. How is this connected to fishes? In a previous study, nine species of fishes that coexist in the small and endangered system of Churince were studied in terms of their diet by Hernández et al. (2017) observing a clear trophic niche differentiation within the Churince. This feeding difference was also reflected in the virus observed within three of those fish species (Taboada et al. 2018), being the invasive specie the one with a wider array of virus, suggesting a less specialized habitat.

Environmental Disturbances and Microbial Niches

The above described microbial interactions in CCB seem to be altered when environmental disturbances change the "trophic game." This was observed first in endemic micro-snails feeding on microbial mats that form oncolites (rounded stromatolites usually formed as a result of the current of a river), where an increment in nutrients, in particular P, drove the snails to smaller size, since they could not digest properly the larger amounts of diatoms that were the result of the fertilization (Elser et al. 2005, 2006) suggesting that the snails live in a "knife-edged" niche: with more P, they die or do not do well, and with less P, they cannot survive. Similarly, changes in the total bacterial communities where observed as a result of manipulated variation in nutrient concentration. Using molecular ecology and metagenomic data in a fertilization experiment in a small pond in the Churince system, the complete community was observed to change after 6 weeks of nutrient input (Lee et al. 2015, 2017).

Bacterial Niches in Cuatro Ciénegas: Concluding Remarks and Some Reflections

Considering its extreme oligotrophy and the different experimental and in situ data described in this and other chapters of this book, we can suggest that the high total bacterial diversity of the Cuatro Ciénegas Basin, as well as the large beta-diversity at very short scales in the aquatic (Espinosa-Asuar et al. 2015) or in the soil (Pajares et al. 2015) communities, can be explained by very narrow but dynamic and spatially heterogeneous Eltonian and Grinnellian niches.

An example of this fine niche microbial partitioning was observed in the growth curves and stoichiometric content of *Bacillus* species that coexist in the same site in CCB (Valdivia-Anistro et al. 2016), where each strain seems to have a slightly different strategy to cope with the oligotrophy, probably driven initially by competition and resulting in coexistence and in many cases codependences.

As a result, the multidimensional space where the theater of life plays in CCB is very rich and full of actors, which shifts constantly, as virus and other process "chop the head" of the dominant species and former "underdogs" of the rare biosphere take their roles (de Anda 2017). Finally, since this play of "life at the edge" has been continued in the same site for a very long time (Moreno-Letelier et al. 2012) with coevolving communities, the "conversations and negotiations" between microscopic actors seem to be the reason why this stage has persisted for such a long time, becoming a true "lost world" where ancient lineages persisted along with their neighbors and the environmental conditions that they have built as geo-engineers that can cycle and recycle all the fundamental elements of life (de Anda 2017).

Acknowledgments This chapter was written during a sabbatical leave of VS in the Department of Ecology, Evolution, and Behavior, University of Minnesota, in Dr. Michael Travisano's laboratory, and of LEE, in the Department of Plant and Microbial Biology, University of Minnesota, in Dr. Peter Tiffin's laboratory, both supported by the program PASPA-DGAPA, UNAM.

References

Aguirre-Von-Wobeser E et al (2014) Two-role model of an interaction network of free-living γ-proteobacteria from an oligotrophic environment. Environ Microbiol 16:1366–1377

Boddy L (2000) Interspecific combative interactions between wood-decaying basidiomycetes. FEMS Microbiol Ecol 31:185–194

Comte L et al (2016) Global test of Eltonian niche conservatism of nonnative freshwater fish species between their native and introduced ranges. Ecography. https://doi.org/10.1111/ecog.02007

Cordero OX et al (2012) Ecological populations of bacteria act as socially cohesive units of antibiotic production and resistance. Science 337:1228–1231

Crespi BJ (2001) The evolution of social behavior in microorganisms. Trends Ecol Evol 16:178–183

Danger M et al (2013) Benthic algae stimulate leaf litter decomposition in detritus-based headwater streams: a case of aquatic priming effect? Ecology 94:1604–1613

De Anda V (2017) MEBS, a software platform to evaluate large (meta)genomic collections according to their metabolic machinery: unraveling the sulfur cycle 2017, GigasciencesGIGA-D-17-00134

de Anda V (2018) ciclo del azufre. metagenomica de la conservaci6n en cuatro cienegas Coahuila. Tesis de Doctorado en Ciencias Biomédicas junio 2018

Elser JJ et al (2005) Response of grazing snails to phosphorus enrichment of modern stromatolitic microbial communities. Freshw Biol 50(11):1826–1835

Elser JJ et al (2006) Early Cambrian food webs on a trophic knife-edge? A hypothesis and preliminary data from a modern stromatolite-based 501 ecosystem. Ecol Lett 9(3):295–303

Espinosa-Asuar L, Escalante AE, Gasca-Pineda J, Jazmin B, Lorena P, Eguiarte LE, Valeria S (2015) Aquatic bacterial assemblage structure in Pozas Azules, Cuatro Ciénegas Basin in México: deterministic vs. stochastic processes. Int Microbiol 18:105–116. https://doi.org/10.2436/20.1501.01.240

Flombaum P et al (2013) Present and future global distributions of the marine Cyanobacteria Prochlorococcus and Synechococcus. Proc Natl Acad Sci 110:9824–9829

Ghorbani R et al (2008) Soil management for sustainable crop disease control: a review. Environ Chem Lett 6:149–162

Gulis V, Suberkropp K (2003) Interactions between stream fungi and bacteria associated with decomposing leaf litter at different levels of nutrient availability. Aquat Microb Ecol 30:149–157

Hernández A et al (2017) Trophic analysis of the fish community in the Ciénega Churince, Cuatro Ciénegas, Coahuila. PeerJ 5:e3637. https://doi.org/10.7717/peerj.3637

Hutchinson GE (1965) The ecological theater and the evolutionary play. New Haven. Yale University Press

Lazdunski AM, Ventre I, Sturgis JN (2004) Regulatory circuits and communication in gram-negative bacteria. Nat Rev Microbiol 2:581–592

Lee Z et al (2015) Stoichiometrically inbalanced shallow ponds in the Cuatro Cienegas Basin is highly sensitive to nutrient enrichement. PLoS One. https://doi.org/10.1371/journal.pone.0123949

Lee Z et al (2017) Nutrient stoichiometry shapes microbial community structure in an evaporitic shallow pond. Front Microbial Ecol 8:949. https://doi.org/10.3389/fmicb.2017.00949 ISSN: 1664-302X. FI 4.165

Martínez-Carranza E et al (2017) Involvement of cyclodipeptides in the competition of bacterial communities in the Churince aquatic system of Cuatro Ciénegas Basin dominated by Gammaproteobacteria. Extremophyles 22:73–85. https://doi.org/10.1007/s00792-017.0978-3

Mille-Lindblom C, Tranvik LJ (2003) Antagonism between bacteria and fungi on decomposing aquatic plant litter. Microb Ecol 45:173–182

Mitri S, Foster RK (2013) The genotypic view of social interactions in microbial communities. Annu Rev Genet 47:247–273

Moreno Fontalba CJ (2016) Ecología de las interacciones entre bacterias en Cuatro Ciénegas, Coahuila, México. Master thesis. Universidad de Sevilla, España. Facultad de Biología

Moreno-Letelier A et al (2012) Divergence and phylogeny of Firmicutes from the Cuatro Cienegas Basin, Mexico. Astrobiology 12(7):674–684

Morón-Ríos A, Gómez-Cornelio S, Ortega-Morales BO, De la Rosa-García S, Partida-Martínez LP, Quintana P, Alayón-Gamboa JA, Cappello-García S, González-Gómez S (2017) Interactions between abundant fungal species influence the fungal community assemblage on limestone. PLoS ONE 12:e0188443

Pajares S et al (2015) Multivariate and phylogenetic analyses to assess the response of bacterial mat communities from an ancient aquatic ecosystem to different scenarios of climate change. Plos One. https://doi.org/10.1371/journal.pone.0119741

Pérez-Gutiérrez RA, López-Ramírez V, Islas A, Alcaraz LD, Hernández-González I, Olivera BC, Santillan M, Eguiarte L, Souza V, Travisano M, Olmedo-Alvarez G (2013) Antagonism influences assembly of a Bacillus guild in a local community and is depicted as a food-chain network. Int Soc Microb Ecol 7(3):487–497

Ponce-Soto GY, Aguirre-von-Wobeser E, Eguiarte LE, Elser JJ, Lee ZMP, Souza V (2015) Enrichment experiment changes microbial interactions in an ultra-oligotrophic environment. Front Microbiol 6:246

Rodríguez-Torres MD, Islas-Robles Á, Gómez-Lunar Z, Delaye L, Hernández-González I, Souza V, Travisano M, Olmedo-Álvarez G (2017) Phenotypic microdiversity and phylogenetic signal analysis of traits related to social interaction in Bacillus spp. from sediment communities. Front Microbiol 8:29

Schuster M, Lostroh CP, Ogi T, Greenberg EP (2003) Identification, timing and signal. 12: specificity of Pseudomonas aeruginosa quorum-controlled genes: a transcriptome analysis. J Bacteriol 185:2066–2079

Soberon J (2007) Grinnellian and Eltonian niches and geographic distribution of species. Ecol Lett 10:1115–1123

Souza V et al (2012) Travel, sex and food: what's speciation got to do with it? Astrobiology 12:634–643

Sridhar KR, Bärlocher F (2000) Initial colonization, nutrient supply, and fungal activity on leaves decaying in streams. Appl Environ Microbiol 66:1114–1119

Stubbendieck RM (2016) Bacterial communities: interactions to scale. Front Microbiol 7:1234

Taboada B et al (2018). The viral structure of the Cuatro Ciénegas Basin, a unique oasis in Northern Mexico, reveals a highly diverse population at a small geographic scale. Appl Environ Microbiol. https://doi.org/10.1128/AEM.00465-18

Valdivia-Anistro J et al (2016) Variability of rRNA operon copy number and growth rate dynamics of Bacillus isolated from an extremely oligotrophic aquatic ecosystem. Front Microbiol 6:1486. https://doi.org/10.3389/fmicb.2015.01486

Velez P, Gasca-Pineda J, Rosique-Gil E, Eguiarte LE, Espinosa-Asuar L, Souza V (2016) Microfungal oasis in an oligotrophic desert: diversity patterns and community structure in three freshwater systems of Cuatro Ciénegas, Mexico. PeerJ 4:e2064

Velez et al (2018) Experimental and molecular approximation to microbial niche: trophic interactions between oribatid mites and microfungi in an oligotrophic freshwater system. PeerJ 6:e5200; https://doi.org/10.7717/peerj.5200

West SA, Diggle SP, Buckling A, Gardner A, Griffin AS (2007) The social lives of microbes. Annu Rev Ecol Evol Syst 38:53–77

Wohl DL, McArthur JV (2001) Aquatic actinomycete-fungal interactions and their effects on organic matter decomposition: a microcosm study. Microb Ecol 42:446–457

Woodward S, Boddy L (2008) Interactions between saprotrophic fungi. In: British mycological society symposia series, vol 28. Academic, New York, pp 125–141

Zapién-Campos R, Olmedo-Álvarez G, Santillán M (2015) Antagonistic interactions are sufficient to explain self-assemblage of bacterial communities in a homogeneous environment: a computational modeling approach. Front Microbiol 6:489

Chapter 7
Microcosms and Mesocosms: A Way to Test the Resilience of Microbial Communities in Cuatro Ciénegas

Nguyen E. López-Lozano, Silvia Pajares, Ana E. Escalante, Luis E. Eguiarte, Valeria Souza, and Gabriela Olmedo-Álvarez

Abstract Given the fragility of the Cuatro Cienegas Basin and the danger of its loss, we started to study the variables affecting the resilience of the microbial community using different experimental approaches. How do microbial communities react to different kinds of perturbations and global change scenarios? We analyzed a series of experimental models that represent different ecosystem compartments: bulk soil, soil crusts, water, and sediment. The experiments were performed in mesocosm or microcosm model systems, which we call in general "cosm" experiments. Different questions were addressed. How does water availability affect the recovery of microbial communities in disturbed soil patches? How do changes in temperature affect microbial crusts? How do bacterioplankton and bacterial mat communities respond to changes in temperature and UV radiation? What would happen to an oligotrophic environment if there was a high nutrient input? Our results suggested perturbations that influenced community structure and community cohesion were stronger in less fluctuating environments. Although it had been suggested that there was a functional (ecological) equivalence between microbial communities,

N. E. López-Lozano (✉)
CONACyT-División de Ciencias Ambientales, Instituto Potosino de Investigación Científica y Tecnológica (IPICYT), San Luis Potosí, Mexico
e-mail: nguyen.lopez@ipicyt.edu.mx

S. Pajares
Unidad de Ecología y Biodiversidad Acuática, Instituto de Ciencias del Mar y Limnología, Universidad Nacional Autónoma de México, Ciudad de México, Mexico

A. E. Escalante
Laboratorio Nacional de Ciencias de la Sostenibilidad (LANCIS), Instituto de Ecología, Universidad Nacional Autónoma de México, Ciudad de México, Mexico

L. E. Eguiarte · V. Souza
Instituto de Ecología, Universidad Nacional Autónoma de México, Ciudad de México, Mexico

G. Olmedo-Álvarez
Departamento de Ingeniería Genética, Cinvestav Unidad Irapuato, Irapuato, Gto, Mexico

V. Souza et al. (eds.), *Cuatro Ciénegas Ecology, Natural History and Microbiology*, Cuatro Ciénegas Basin: An Endangered Hyperdiverse Oasis,
https://doi.org/10.1007/978-3-319-93423-5_7

our results on N_2-fixing microorganisms of two arid ecosystems showed functional differences, even though similar species occur in both systems. Currently, new experiments are being carried out in "cosms" with replicates of the sediment and water interphase. Sadly, in one of the sites we have studied better, the Churince ecosystem, its conditions were dire the last time that sediment and water were sampled there. This ecosystem has now nearly disappeared. Nevertheless, the sediment in these fish tanks seems to be recovering its original structure. While seeming like a white rhino in a zoo, it might be the only ecosystem of this kind left to learn about what we lost.

Introduction

Earth's climatic change and human activities are modifying the biota and physical characteristic of the planet at an accelerating pace. Predicting their ecological consequences is complex, because there are many factors that influence the functioning of an ecosystem (Grimm et al. 2013). Cuatro Ciénegas Basin (CCB) is a perfect example of this complexity, because this valley possesses a particular geological history that makes its territory atypical and highly heterogeneous. In general, the biota is extraordinarily diverse. This valley was recognized in recent years to be a fragile oasis subjected to climatic change, where excessive water extraction, pollution by tourism, settlements and agriculture, and inadequate agricultural practices have worsened its condition. CCB faces challenges, such as natural fluctuations in water availability and alterations to hydrology and biogeochemistry which interacts with other stressors, like eutrophication and environmental warming. Regarding human activities and climate change, we need to obtain pertinent and mechanistic ecological information in order to improve our ability to predict future scenarios of CCB ecology.

Microorganisms constitute the largest proportion of the world's biodiversity, play key roles in biogeochemical cycles, and are a fundamental component of ecosystems (Fuhrman 2009; Whitman et al. 1998). Therefore, understanding the major factors that drive changes in the diversity of microbial communities is important for predicting ecosystem responses to a changing environment. However, knowledge is still limited regarding the underlying mechanisms of microbial behavior, community assembly, and functional consequences of diversity shifts (Widder et al. 2016). Despite the low number of ecological microbial studies using experimental approaches, they are promising as ecologists are aware of the need for robust experimental designs to address microbial ecological issues (Jessup et al. 2005; Strickland et al. 2009).

The difficulty in dealing with a part of and the whole scenario at the same time is widely recognized in science. Temporal and spatial scales of environmental problems are so broad that a mechanistic understanding of cause-effect relationships is difficult to gain, especially with traditional correlational studies. To bridge the gap, a more sophisticated approach is needed to integrate novel experimental methods that can support more realistic levels of complexity (Odum 1983). Thus, replicate experiments under controlled conditions are a good way to reveal causality and the underlying mechanisms of ecological changes. The simplification of controlled experiments—controlled variables and replicas—is necessary to address questions that are inaccessible through field observation or correlation (Jessup et al. 2004). In this sense, model systems—simplified representations of more complex systems—are simplifications of the study systems, especially when they are conducted on small spatial and temporal scales, such as in microcosms and mesocosms (Stewart et al. 2013). Model systems should have three useful features: tractability, generality, and realism (Levins 1966).

Mesocosms and Microcosms: Controlled Experiments to Look for Ecological Explanations

The terms mesocosms and microcosms are often (but not always) used to define in situ and in vitro laboratory experiments that gain an understanding of ecological processes. Draggan and Reisa (1980) defined microcosms as "small experimental units designed to contain important components and to exhibit important processes occurring in a whole ecosystem." The term mesocosm was proposed for middle-sized experiments falling between laboratory microcosms and the large, complex, real world macrocosms that include more biological complexity (Grice and Reeve 1982a, b; Odum 1983). Other definitions describe microcosms as generic systems in which species composition and abiotic characteristics are determined by the operator. Mesocosms are more natural, semirealistic systems with an indigenous combination of organisms and abiotic conditions (Wijngaarden et al. 2005). The prefix micro-, meso-, or even macro- may be used to define the size of such an experimental unit. However, these limits are arbitrary when we talk about microorganisms where the formal categorization is not so rigid.

The realism these systems offer is most important (Bloesch 1988). It is evident that "cosms" do not exactly mimic natural systems at all organizational levels, because they are limited in size, time, and mass of both biotic and abiotic components. In addition, "cosms" have boundaries that restrict interaction with the rest of the ecosystem. To be useful, "cosms" should minimize variability without reducing realism, and to not compromise reproducibility for ecological realism, the size and complexity of "cosms" depend on the nature of the question to be answered or the hypothesis to be tested (Wijngaarden et al. 2005).

Resilience and Functional Equivalence of Microbial Communities

Because of their large population sizes and short generation times, microbial community turnover is faster as well as easier to observe and replicate than that of macroscopic organisms. Microbial experimental systems thus provide an important between the theory and the complexity of nature. Additionally, microbial communities are particularly appropriate to study how natural communities change over time (Fierer et al. 2010). These systems can deal with ecological questions over "long temporal" scales, an important requirement for studying community-level responses to environmental changes (Jessup et al. 2004).

The recent drastic increases in extinction rates of plant and animal species have made ecologists consider the consequences of biodiversity loss, because biodiversity supplies economically valuable ecosystem goods and services on which humanity depends. Although most biodiversity and conservation research has focused on large organisms, empirical and experimental evidence have shown that changes in microbial communities caused by disturbance may directly affect ecosystem functioning (Allison and Martiny 2008). Consequently, we must understand the links among composition, physiological traits, and disturbance responses in microbial communities.

Given the high diversity and traits of microbial communities, microorganisms are thought to be resistant, resilient, and functionally redundant. Some microbial groups possess a high degree of metabolic flexibility and stress tolerance, suggesting that microbial communities might retain their composition and function after environmental fluctuations. However, few studies have observed a resistant microbial composition (Bowen et al. 2011), and data often suggest microbial sensitivity to disturbance (Allison and Martiny 2008).

Assuming that disturbance changes community composition, communities appear to return to their original composition and function after a while. This phenomenon is known as resilience and is possibly due to peculiarities found in microorganisms, such as high abundance, widespread dispersal, potential for rapid growth rates, and rapid evolutionary adaptation through horizontal gene transfer that allows sensitive microorganisms to adapt to new environmental conditions. The key point of resilience is the length of time between the disturbance and when the community returns to its original composition. The literature suggests, as found in this chapter, that microbial composition usually differs from that of undisturbed communities for a time scale of at least a few years (Baho et al. 2012; Allison and Martiny 2008).

Although resilience may occur, the community composition could remain in an altered state after a perturbation if the disturbance intensity is too high and frequent. However, communities maintaining their function are likely due to a high degree of functional redundancy among microbial communities (Baho et al. 2012). In fact, functional redundancy in microbial communities is considered to be high with many species performing the same functions (Wohl et al. 2004; Langenheder et al. 2005; Comte and del Giorgio 2010; Baho et al. 2012). Nevertheless, functional redundancy is difficult to establish, as it requires detailed information about microbial identity, microbial traits, and microbial ecology to perform a specific process. Despite the considerable emerging information obtained by metagenomics, this topic is still new to microbial ecologists.

In order to address resilience velocity and functional redundancy that are crucial to maintaining ecosystem processes, one needs to manipulate a microbial composition in such a way that enables recognition of process rate differences that are attributed directly to compositional differences and not simply to physiological responses of the original community to the natural fluctuations in environmental conditions. It is more feasible to manipulate microbial community composition in semicontrolled conditions than in the open field. "Cosms" seem to be a true asset for assessing environmental risks, since they can be used to test the disturbance effect in realistic scenarios without endangering the original ecosystem. One of the primary goals of this type of study should be to place ecological risk analysis in the broader context of regional natural resource management, a central theme of applied ecology (Boyle and Fairchild 1997). A principal question that "cosms" could answer is as follows: What is the expected magnitude of the disturbance effect in ecosystems with different characteristics reflective of regional ecological conditions?

Cuatro Ciénegas "Cosm" Experiments

In light of CCB different environmental stressors, and the recent disturbances that it is suffering, the main objective of this chapter is to review the experiments performed using "cosms" to test the resilience of the microbial communities in its different ecosystems. In this section we will show how CCB is indeed a fragile oasis, despite having previously survived for a very long time. With a series of experiments, the functions that microbes perform in the different ecosystems became evident as well as the danger of the potential loss of its remarkable microbial diversity.

"Cosm" Experiments with Soil Bacteria

Experimental Studies of the Succession in Soil Bacteria in Different Humidity and Nutrient Conditions

Soil is a vital part of the natural environment and provides a habitat for a wide range of organisms. Soil characteristics influence the species distribution that inhabit it. Soil acts as a filtration system for surface water and controls the chemical flow between the atmosphere and the Earth. It also acts as carbon storage and maintenance of atmospheric gases. Soil constantly changes and develops through time. It responds to environmental changes and human land use. Soils not only reflect natural processes but also record present and past human activities. Some changes are short in duration and reversible, but others are permanent fingerprints on soil development that are reflected in the biotic community composition.

Changes in community composition through time are known as succession. In ecological theory, succession is defined as the (more or less) predictable manner by which communities change over time during the colonization of a new environment, or following a disturbance—primary and secondary succession, respectively (Begon et al. 2006).

The gypsum-based soil system at CCB is one of the most oligotrophic environments in the world, but it surprisingly possesses a high microbial diversity in comparison to other arid soils (Lopez-Lozano et al. 2012). Arid soils are considered to be vulnerable to human activities under most global climate change scenarios (Intergovernmental Panel on Climate Change 2015). Due to recent and ongoing overexploitation of the deep aquifer in the CCB, we were particularly interested in evaluating the microbial community sensitivity of soils after two cases of disturbance in humidity. Lopez-Lozano et al. (2013) evaluated microbial biodiversity recovery and biogeochemical characteristics of the Churince within the CCB during microbial secondary succession over a 1-year period after disturbance. Their approach was designed to be experimentally analogous to forest sections cleared by removing the biota of an area and tracking recolonization by migration.

The Churince system in CCB consists of a spring, an intermediate lagoon, and a desiccation lagoon connected by short shallow creeks. Lopez-Lozano et al. (2013) selected two sites 1 kilometer apart from each other. One site bordered a small river creek (named River) with ca. 60% plant coverage dominated by the grass *Sporobolus airoides*. The other site (named Dry Lagoon) was closer to the desiccation lagoon, with only ca. 10% plant coverage and dominated by the gypsophile plant *Sesuvium erectum*. Both sites had alkaline soils with little variation in pH (8.5–8.8). In January 2007, 20 kg of soil were collected from each site. The soil was mixed and used to construct microcosms of 1 kg of soil in permeable mesh nylon bags (25 × 25 cm). The microcosms were sterilized by autoclaving followed by a 25 kGy dose of gamma rays to eliminate all biota. In February 2007, 8 × 8 m plots were established at each site. The plots were divided into 64 quadrants of 1 m^2, and 40 microcosm bags were placed randomly within the plots as replicates. Every 3 months, three microcosm bags were randomly collected for 1 year (3, 6, 9, and 12 months) for a total of 12 bags in each site. In addition, three samples from 15 cm^3 of undisturbed soil next to the microcosms in each plot were randomly collected at the beginning and at the end (12 months) of the experiment. These samples were collected not only to describe the community that potentially could colonize the microcosms but also to compare the recovery of the microcosms with the undisturbed soil.

To understand the relationships between the microbial community composition changes and nutrient transformations that occurred during the secondary succession process in the Churince soils, the soil samples were biogeochemically characterized and compared. The CCB soil at the Dry Lagoon site was more oligotrophic (C/N/P ratio of 125:5:1) than the River site (C/N/P ratio 300:16:1) and in comparison with a general “average” soil C/N/P of 186:13:1 (Cleveland and Liptzin 2007). In general, there were consistent differences in the soil nutrients between sites across time (Tables 7.1 and 7.2). Four factors were observed to strongly contribute to the differences in our sites: humidity, total N, total organic C, and total P. The first three factors were identified as the most correlated variables with the relative abundance of the bacterial groups. These factors are considered to be some of the best soil variables for predicting microbial community composition in arid lands (Collins et al. 2008; Wall and Virginia 1999). Total phosphorus in the soil explained some vari-

Table 7.1 Biogeochemical parameters of two sites were "cosms" microbial succession was studied in the Churince system in Cuatro Ciénegas Basin (CCB), before the experiment and at two latter times, and statistical significance of the differences

Site	Dry Lagoon			River			F
Time	Sterilized soil	6 months	12 months	Sterilized soil	6 months	12 months	Site*time
Organic Carbon (total)	0.3	2.5	2.9	6.5	7	4	7.7**
Nitrogen (Total)	0.15	0.13	0.24	0.8	0.6	0.7	0.7
Phosphorus (Total)	0.02	0.01	0.02	0.04	0.03	0.04	3.91*
Organic C (disolved)	508	218	13.5	239	371	94	0.36 ns
Organic N (disolved)	4.6	2.1	8.2	27	3.7	10.3	4.5*
Organoc P (dissolved)	0.02	1.8	1.8	1.5	1.4	1.4	2.5 ns
Inorganic P (dissolved)	0.02	0.02	0.005	0.007	0.006	0.004	10.6***
Microbial C	0	267	92	0	348	219	0.89 ns
Microbial N	0	1.6	2	0	5.3	5.5	9.3**

F values and significance levels; $^{*}P < 0.05$, $^{**}P < 0.01$, $^{***}P < 0.0001$, *ns* not significant

ability as expected, because phosphorus was the most limiting nutrient in the CCB (Elser et al. 2005).

The community structure in each of the microcosm bags was evaluated by 16S rDNA 454 pyrosequencing of amplicons. The majority of sequences (>80%) belonged to one of the nine major phyla commonly found in soils (arranged in order of abundance): *Acidobacteria*, *Proteobacteria*, *Bacteroidetes*, *Firmicutes*, *Chloroflexi*, *Deinococcus-Thermus*, *Gemmatimonadetes*, *Actinobacteria*, and *Candidatus* division TM6 (Fig. 7.1). Even with the higher sequencing effort afforded by 454 pyrosequencing, it was estimated that only 52–81% of the microbial communities in the samples was captured. In more fertile and wet soils analyzed with similar sequencing effort, the maximum number of OTUs at 97% identity seldom exceeded 5,600 OTUs (Roesch et al. 2007). Our most diverse sample (River) had an estimated 9000 OTUs at 97% identity. Despite the nutrient content differences observed between sites, 16S rDNA sequence analysis showed similar variations in diversity and community composition in both sites. The Shannon index—overall diversity—was higher in the River location during the entire sampling period. At both sites, the diversity increased in the third and sixth months, decreased in the ninth month, and then increased again in the twelfth month following a similar fluctuating biomass pattern. Comparing the diversity and adjacent undisturbed soils observed in the last sampling date, the undisturbed soil had a higher diversity than the microcosm samples.

Table 7.2 Soil biogeochemical parameters in each sampling date for both sites Dry Lagoon and River

Site	Dry Lagoon					River					Effect		
											Site	Time	site*time
Time	Sterilized soil	3 months	6 months	9 months	12 months	Sterilized soil	3 months	6 months	9 months	12 months	*F* values		
TOC (mgg⁻¹)	0.3 ± 0.13	1.8 ± 0.60	2.5 ± 0.03	3.9 = 0.30	2.9 ± 0.20	6.5 ± 0.60	9.8 ± 0.86	7 ± 1.10	15 ± 2.10	4 ± 1.70	61.4***	10.68***	7.67**
TN (mgg⁻¹)	0.15 ± 0.01	0.16 ± 0.06	0.13 ± 0.004	0.23 ± 0.01	0.24 ± 0.02	0.8 ± 0.01	0.6 ± 0.08	0.6 ± 0.03	0.8 ± 0.04	0.7 ± 0.06	226.9***	3.58*	0.77 ns
TP (mgg⁻¹)	0.02 ± 0.001	0.01 ± 0.002	0.01 ± 0.001	0.05 ± 0.002	0.02 ± 0.001	0.04 ± 0.006	0.08 ± 0.045	0.03 ± 0.001	0.15 ± 0.001	0.04 ± 0.001	19.76**	9.37**	3.91*
DOC (µgg⁻¹)	508 ± 53	817.4 ± 146	218.4 ± 26	170 ± 206.6	13.5 ± 14	239 ± 33	712 ± 132	371 ± 27	830 ± 375	94.3 ± 47	3.75 ns	4.66*	0.36 ns
DON (µgg⁻¹)	4.6 ± 1	6.7 ± 0.9	2.1 ± 0.5	8.2 ± 1.1	8.2 ± 4.5	27 ± 1	10.3 ± 1.7	3.7 ± 0.6	25.5 ± 2.2	10.3 ± 0.6	13.09**	12.72**	4.47*
DIP (µgg⁻¹)	0.02 ± 0.004	0.004 ± 0.004	0.02 ± 0.006	0.04 ± 0.006	0.005 ± 0.003	0.007 ± 0.004	0.01 ± 0.008	0.006 ± 0.006	0.14 ± 0.030	0.004 ± 0.04	8.95**	24.81***	10.57***
DOP (µgg⁻¹)	1.6 ± 0.1	2.3 ± 0.2	1.8 ± 0.01	2.1 ± 0.6	1.8 ± 0.06	1.5 ± 0.05	1.5 ± 0.05	1.45 ± 0.05	0.65 ± 0.1	1.4 ± 0.04	1.33 ns	15.83**	2.48 ns
Ammonium (µgg⁻¹)	0.4 ± 0.1	0.08 ± 0.08	2.9 ± 0.8	2 ± 0.9	1 ± 0.2	4.8 ± 0.2	4.3 ± 1.5	6.9 ± 1.2	7.8 ± 0.9	0.6 ± 0.1	29.08***	10.68**	4.31*
Nitrates (µgg⁻¹)	0 ± 0	0.6 ± 0.5	0.3 ± 0.2	1.1 ± 0.2	0.3 ± 0.2	1.4 ± 0.7	0 ± 0	0.2 ± 0.2	0 ± 0	0 ± 0	13.91**	0.96 ns	1.61 ns
micC (µgg⁻¹)	0 ± 0	172.3 ± 45	267 ± 127	62 ± 10	92 ± 26	0 ± 0	182 ± 12	348.6 ± 69	139 ± 18	219 ± 38	10.27**	6.22**	0.89 ns
micN (µgg⁻¹)	0 ± 0	3.4 ± 0.6	1.6 ± 0.4	7.3 ± 0.2	2 ± 0.3	0 ± 0	3.2 ± 1	5.3 ± 1.4	6.8 ± 2.8	5.5 ± 0.5	2.89 ns	3.69*	9.31**
C:N	2	11.25	19.23	16.96	12.08	8.12	16.33	11.66	18.75	5.71	–	–	–
micC:micN	–	50.67	166.87	8.49	46	–	56.87	65.77	20.44	39.81	–	–	–

Analysis was done as a 2 way ANOVA, the mean ± SD with *F* values and significance levels; $^{*}P < 0.05$, $^{**}P < 0.01$, $^{***}P < 0.0001$, *ns* not significant

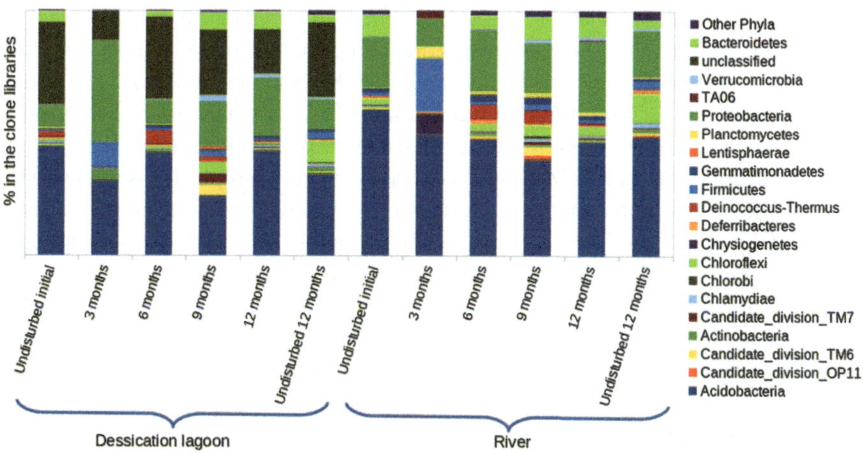

Fig. 7.1 Taxonomic distribution of the 16S rDNA library sequences at the phylum level of the soil bacteria from the two experimental "cosms" sites at the Churince system in the CCB. Data from all sampling dates (3, 6, 9 and 12 months), the undisturbed soil at the beginning of the experiment (before the sterilization) and the last sampling date (12 months undisturbed) are shown

What is the meaning of the observed microbial diversity quantified as a function of 16S rDNA genes? We wanted to interpret this diversity in terms of ecological function. For this reason, community composition was evaluated with "ecological groups" formed by the dominant bacterial "species" that correlated with high or low C mineralization rates (copiotrophs or oligotrophs, respectively) (Fierer et al. 2007, 2010) or known autotrophs (i.e., phototrophs, such as *Cyanobacteria* and *Chloroflexi*). In other cases, some metabolic capabilities were deduced from the relatedness of recovered 16S rDNA to their genetically close cultivated organisms. Despite the broad ecological classification based only in 16S rDNA gene sequences, the results showed interesting trends in oligotrophs, copiotrophs (organism that need more nutrients), heterotrophs, and autotrophs.

16S rDNA libraries obtained from samples in both sites were dominated by *Acidobacteria*. This is one of the most common phyla found in soil libraries worldwide (Janssen 2006). *Acidobacteria* are generally oligotrophic (Eichorst et al. 2007; Fierer et al. 2007). They comprise 50% of the clone libraries in arid soils (Dunbar et al. 1999; Kuske et al. 1997) but are less abundant in more nutrient-rich agricultural soils (Nagy et al. 2005; Roesch et al. 2007). Hence, it was not surprising that *Acidobacteria* represented 30–60% of the bacteria in all DNA libraries in the CCB sampling sites.

The generally copiotrophic *Bacteroidetes* (Fierer et al. 2007) were more abundant in the 6-month sample. These phyla possibly appeared when organic material accumulation was high enough to sustain the mineralization rate of this group. There was a decrease in dissolved organic carbon between 6 and 9 months, while the microbial carbon increased in the soil microcosms.

Autotrophic groups, such as *Cyanobacteria*, were abundant at both sites. This phylum is common in most environments, including soil and the biological soil crust of arid zones (Gundlapally and Garcia-Pichel 2006; Nagy et al. 2005). The abundance of yet another autotrophic group, *Chloroflexi*, increased during the initial stages of succession (3–9 months) but decreased in the last sampling date (12 months).

The majority of the colonizers were opportunistic heterotrophs (mostly members of *Firmicutes*, *Betaproteobacteria*, and *Gammaproteobacteria*), with the exception of high *Chloroflexi* concentrations in the River sample during the first samplings. Especially in the Dry Lagoon site, the initial colonizers possibly benefited from the dissolved organic carbon released during the sterilization process. The nutrient pulse could have facilitated heterotroph colonization that benefited from available resources in the disrupted microcosm. A similar result was observed in a laboratory microcosm experiment carried out with Churince sediment samples (unpublished data).

Preferential growth of opportunistic heterotrophs during early succession agreed with the findings from another study using outdoor sterile microcosms seeded by rainwater bacteria (Langenheder and Szekely 2011). These authors found that neutral and species sorting processes interacted during bacterial community assembly. The importance of each process depended on how many generalists and specialists were present in the community.

In the microcosms of our soil study, we believe that the initial faster-growing community (generalists) depleted the nutrients but then was outcompeted by groups that had alternative energy sources (specialists). At 6 months, groups with more specialized metabolic capabilities increased, and the low abundant, pre-sterilization taxa or "the rare biota" that were not present after 3 months returned to the community. Measured with the Bray-Curtis distance and Jaccard coefficient using OTUs at 97% identity, samples at both sites achieved a peak 3 months after sterilization, while it was lower before sterilization as well as 6 and 9 months after sterilization. After a year, the microcosm communities were more similar to the undisturbed soil communities at each site.

While each site differed in both biogeochemistry and biodiversity, both exhibited similar community composition at the beginning of the experiment. However, successional patterns differed in later stages. This finding suggests that a species sorting effect occurred at the beginning of the experiment, followed by a more neutral process in later stages. In time, the microcosm communities begin to be more similar to the communities from undisturbed surrounding soil, but they do not reach the same composition.

In addition, phylogenetic and ecological group analyses suggested that the communities responded in a similar trajectory, with initial colonization by heterotrophic generalists and later colonization by specialists. However, the River site was characterized by higher nutrient concentration and diversity along with the continuous presence of highly abundant *Chloroflexi*, suggesting an early food web based on primary production. This pattern is similar to the early stages of the autotrophic succession suggested by Fierer et al. (2010).

In our study, the community composition of the soil microcosms after perturbation did not recover to similar communities found in the undisturbed soil community after 1 year. Assuming that there is not only one "climax" community in terms of composition, a community "recovered" when it had performing conditions similar to the neighboring undisturbed community. The community performance can be inferred by the soil nutrient content and characteristics.

Due to the significant differences in physicochemical parameters between microcosms and control sites at 12 months, Lopez-Lozano et al. (2013) concluded that the small patches (1 kg mesh bags) did not recover to resemble undisturbed soil in either site after a year of migration and succession despite Churince soil diversity. The sites followed parallel paths of recovery at the beginning of succession and diverged in later stages. Interestingly, the succession of small, perturbed spots is very slow, revealing that the temporal scale is important for this community in terms of resilience. Long-term monitoring is necessary to better understand the temporal patterns and natural variability of this area. Microbial community changes due to disturbance may directly affect ecosystem processes that are vital to a protected natural area threatened by overexploitation of aquifers as is the case for the CCB.

Biological Soil Crusts: Community Structure and Activity of N_2-Fixing Microorganisms of Two Arid Ecosystems

Biological soil crusts (BSC) are complex microbial communities composed of photosynthetic and nitrogen-fixing bacteria, algae, fungi, moss, and lichens that are abundant in arid ecosystems, including deserts (Belnap et al. 2004; Eldridge 2000; Nunes da Rocha et al. 2015; Arroyo et al. 2015). These microbial communities develop within the first 5 cm of the soil surface, mainly between vegetation patches (Steven et al. 2013; Maier et al. 2014). In North America, BSC occupy up to 70% of the soil surface (Yeager et al. 2004; Johnson et al. 2012a), and their microbial components are considered to be the main drivers of key ecosystem processes, such as carbon and nitrogen fixation in arid soils.

Microbial composition of BSC varies as a function of topography, seasonality, soil type, and the spatial scale of the study (Makhalanyane et al. 2013; Steven et al. 2013). Microbial organisms of BSC are assumed to be adapted to water pulses associated with sparse and unpredictable rain events, elevated evaporation rates, and long periods of UV light exposure. For example, cold and hot North American deserts differ in BSC composition with lichens dominating cold deserts and *Cyanobacteria* dominating hot deserts (Belnap et al. 2001). Some *Cyanobacteria* are better adapted to either cold or hot deserts. The Cyanobacterium *Microcoleus vaginatus* is the dominant species in cold deserts, while *Microcoleus steenstupii* is dominant in hot deserts (Garcia-Pichel et al. 2013). These two *Cyanobacteria* groups function better depending on their native ecosystem temperatures (Garcia-Pichel et al. 2013), suggesting different responses might occur (with potential impacts at the ecosystem level) with climate change based on the BSC composition.

Microbial communities mediate many ecosystem processes and changes in their functionality may have a global impact. Changes in community microbial composition are irrelevant to the functional redundancy found in different microbial groups, although few studies have explicitly tested this assumption (Wohl et al. 2004; Baho et al. 2012). We investigated the functional (ecological) equivalence of the community structure and activity of N_2-fixing microorganisms of two arid ecosystems (Chihuahuan desert, Cuatro Ciénegas Basis (CCB) and Sonoran Desert, Valle de Guadalupe) with contrasting temperature-precipitation regimes (summer and winter rain, respectively) (unpublished results).

The hypothesis to be tested was whether the BSC microbial community composition was influenced by the functional rates of ecosystem processes. To test this hypothesis, we designed microcosm-controlled reciprocal transplant experiment in the laboratory and determined the community composition through a culture-independent approach by fingerprinting *nifH,* the gene involved in nitrogen fixation, for gene diversity through TRFLP analysis. We also estimated the N_2-fixing potential through an acetylene reduction assay (ARA) at two contrasting temperatures (15 and 30 °C) for both communities (Fig. 7.2). We used two-way ANOVA to test for the effects of sample origin (for determining microbial composition), incubating environment and their interaction to functional response (N_2-fixing potential).

Two main conclusions were obtained from the microcosm BSC's experiments: (1) BSC composition was specific for the site of origin, and (2) changes in BSC composition affected their functional traits. We found significant differences between communities both in composition and functional measures.

Regarding composition, the BSG samples from the different sampling sites could be clearly identified that allowed us to use the "site of origin" as a way to determine compositional differences. Site of origin differences were obvious, but neither the incubation temperature nor the sampling times seemed to change the composition in a clear direction. This observation was particularly evident at the beginning and at the end of the 180 days experiment. This pattern of similarity or beta-diversity suggests that the community assembly process of the studied BSGs is the result of historical factors (i.e., dispersion limitation, stocasticity, historical environmental conditions) and current environmental factors (Martiny et al. 2006; Lindström and Langenheder 2012; Nemergut et al. 2013).

Regarding an environmental factor considered to be a perturbation (i.e., temperature), the BSG samples differed significantly in terms of their site of origin, but not in terms of the experimental incubation temperature, particularly at the beginning and end of the experiment. These findings were probably due to an adaptation process during the intermediate time periods (Reed and Martiny 2013). Finally, these results were consistent with suggesting the occurrence of local adaptations and biogeographical patterns in microorganisms (Martiny et al. 2006).

The results of this experiment indicated that BSG microbial communities with different phylogenetic compositions responded differently even under the same environmental conditions. These results also provided evidence that microbial community composition was the result of historic selective pressures at the site of origin (Strickland et al. 2009) and that functional equivalence could not be assumed even when groups, such as *Cyanobacteria*, were present in different biological soil crust communities. Overall, the results from these experiments contradicted the assumption that microbial communities in the BSC—nitrogen-fixing bacteria, in particu-

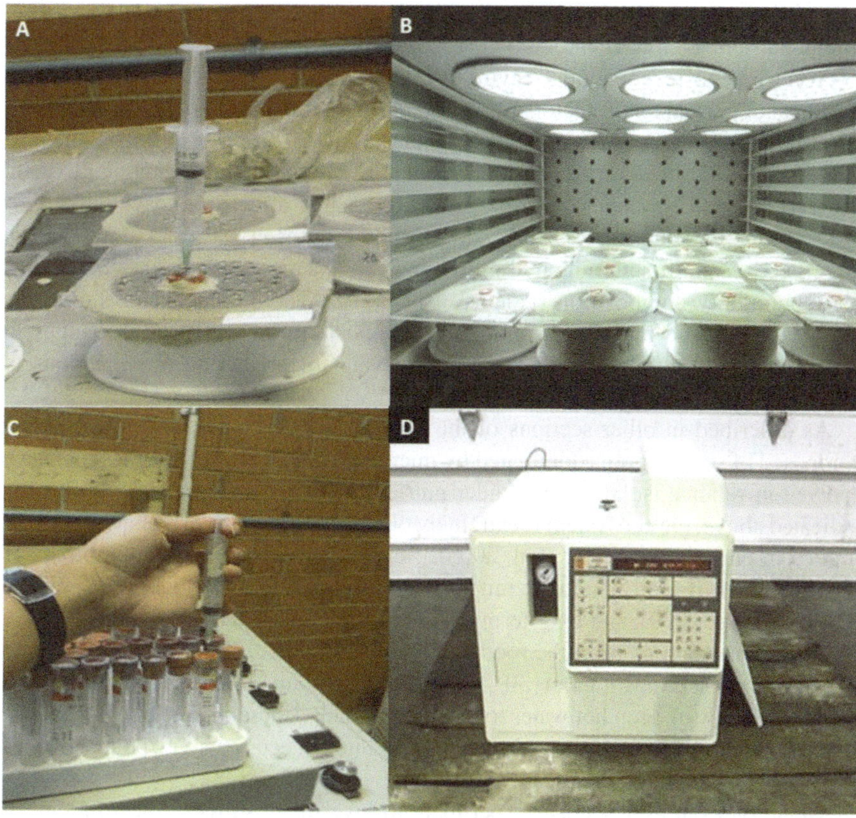

Fig. 7.2 Mesocosms and Acetylene Reduction Assay (ARA). (**a**) Plates with samples of Biological Soil Crusts (BSC) were set up on top with an acrylic cap and vacutainer rubber stopper that sealed the air and allowed the quantification of changes in the composition of injected acetylene (10% of total air volume); (**b**) Sealed samples in an incubator with controls for temperature, humidity and light; (**c**) Gas sampling into vacutainers; (**d**) Gas chromatographer to measure acetylene: ethylene ratio after 72 h of incubation

lar—exhibited functional redundancy, resistance, and resilience in the face of environmental perturbations (Strickland et al. 2009; Reed and Martiny 2013;; Zelikova et al. 2012).

"Cosm" Experiments with Aquatic and Mat Bacteria and Global Change Caused by Humans

In oligotrophic aquatic systems, changes in environmental conditions will affect ecosystem functioning by modifying aquatic bacterial community composition and its associated metabolic capabilities (Fuhrman 2009). These communities will play a key role in the ecosystem response to environmental change due to their central role in nutrient cycling (Falkowski et al. 2008).

According to recent global environmental change scenarios, two of the most significant environmental factors affecting aquatic ecosystems are temperature and UV irradiance (Daufresne et al. 2009; Häder et al. 2011). UV light is known to induce oxidative stress in aquatic bacteria and to produce a cell-damaging oxygen reactive species (Maranger et al. 2002) that ultimately affects photosynthesis, biomass production, and phylogenetic composition (Santos et al. 2012). Temperature affects ecosystem functioning by limiting metabolic activity that defines the microbial community composition (Adams et al. 2010). Therefore, understanding the changes in aquatic microbial community composition in response to changes in temperature and ultraviolet irradiation is relevant for predicting modifications in aquatic systems.

As described in other sections of the book, the CCB is an ancient ecosystem composed of several pools dominated by microbial mats and stromatolites that have evolved in relative isolation and under nutrient constraints (Elser et al. 2006). As indicated above, the ecosystem is extremely oligotrophic and in particular is severely limited in phosphorus (Elser et al. 2005). Nevertheless, the aquatic microbiota has high levels of diversity and endemism (Souza et al. 2008), complex metabolisms (Breitbart et al. 2009), and long-term population stability (Souza et al. 2012). These bacterial communities form the food web base (Souza et al. 2006), inhabit slightly warm thermal springs (about 30 °C) with a constant environment due to the continuous flow of deep hot water to the springs. Thus, these unique microbial mats are suitable model systems to evaluate how this ancient ecosystem responds to long-term environmental disturbances.

We carried out a mesocosm experiment to investigate shifts in the structure of bacterioplankton and bacterial mat communities from the CCB in response to changes in temperature and UV radiation. In July 2009, we constructed 15 mesocosms in 40 L sterilized tanks with composite water samples from three small adjacent natural pools. In order to replicate these complex microbial mat communities without perturbing their natural environment, 2 years prior to the mesocosm experiment, we placed acrylic trays with frosted glass slides in the pools to seed microbial mat-forming communities ("microbial mat catchers").

The mesocosm experiment consisted of the following five treatments, each replicated three times: (a) control tanks with a constant temperature of 30 °C and a UV intensity similar to that of the original pools, (b) fluctuation tanks with temperature shifts (26–43 °C in summer and 0–27 °C in winter) and a UV intensity similar to that of the pools, (c) 40 °C tanks with a constant temperature of 40 °C and a UV intensity similar to that of the pools, (d) UV plus tanks with artificially increased UV radiation for 12 h/day and a temperature similar to that of the pools, and (e) UV minimum tanks with reduced UV light via acrylic filters and a temperature similar to that of the pools. The mesocosms were randomly placed in a well-illuminated and aerated room at the Centro Bachillerato Tecnológico Agropecuario (CBTA22) in the village of Cuatro Ciénegas. In order to observe the changes in biodiversity, we obtained 16S rDNA gene clone libraries from the bacterioplankton at different time

Fig. 7.3 Mesocosm experiment with "synthetic" microbial mats from CCB. (**a**) Poza Azul; (**b**) "microbial mat catchers" in the natural pools; (**c**) mesocosm experiment

periods, including short term (40 days = Ts), medium term (4 months = Tm), and long term (8 months = Tf), as well as from the microbial mats at the end of the experiment (8 months = Mat-Tf).

The original composition of the bacterioplankton community (T0) was dominated by *Actinobacteria*, followed by *Proteobacteria* (mainly *Alpha*, *Beta*, and *Gamma* classes) and *Bacteroidetes*, while mesocosm communities were largely dominated by *Alphaproteobacteria* (Fig. 7.3; Pajares et al. 2012, Pajares et al. 2013). Each treatment had a distinct shift in its bacterial composition during the experiment, indicating a community response to the different experimental environmental condition: phylogenetic bacterioplankton composition and structure exhibited drastic changes in response to fluctuating temperature (fluctuating tanks), increased UV radiation (UVplus tanks), and constant temperature (40 °C tanks). The long-term impact of these environmental stresses reduced diversity and is selected for generalist aquatic bacterial populations, such as *Porphyrobacter* (*Alphaproteobacteria*) (Pajares et al. 2013).

Fluctuating temperature induced the largest decrease of bacterial richness and phylogenetic diversity during the experiment, indicating that frequent temperature changes reduced the abundance of several species, especially autotrophs. In contrast, there was an increase in bacterioplankton diversity over time in the constant temperature (40 °C tanks) and UV protection (UV minimum tanks) treatments, suggesting that stable environments promoted the establishment of complex and diverse bacterial communities in these pools (Pajares et al. 2013). Interestingly, the bacterioplankton structure changed in the short-term experiment (Pajares et al. 2012), most likely due to key community members having slower-growing rates (i.e., *Cyanobacteria*). This finding highlights the importance of following the compositional changes of key taxonomic groups in long-term experiments.

At the end of these mesocosm experiments, 16S rDNA sequence analysis was also performed on the bacterial mat communities (Mat-Tf) retrieved both from the natural pools and the mesocosms. The sequences were mainly from *Cyanobacteria* and *Alphaproteobacteria* lineages (Fig. 7.4; Pajares et al. 2015) which was consistent with previously reported findings for microbial mats from the same pools

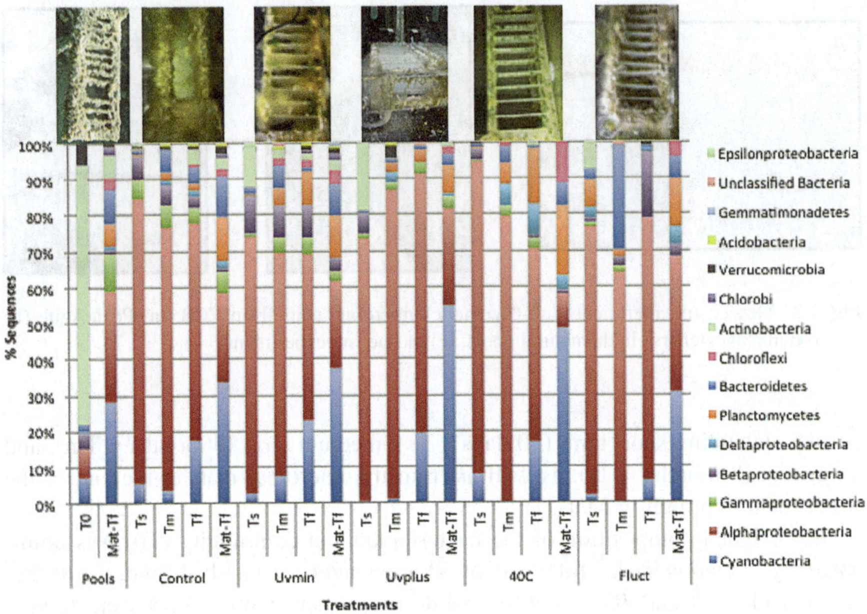

Fig. 7.4 Relative abundance of bacterial groups (measured with 16S rDNA genes) in the different treatment mesocosms using water and strains originally form CCB. (**a**) Bacterioplankton in the pools at the beginning (T_0), short-term (Ts), medium-term (Tm), and at the end of the experiment (Tf); (**b**) microbial mats at the end of the experiment (Mat-Tf) in the pools and treatments. Bars in the photos above represent the "microbial mat catchers" retrieved from each environment

(Bonilla-Rosso et al. 2012) and for stromatolites from Cuatro Ciénegas (Breitbart et al. 2009).

These findings indicated that bacterial communities trapped in the "microbial mat catchers" from the natural pools and the experimental treatments contained microbiota representative of the original ecosystem. Furthermore, a large proportion of OTUs from the microbial mats had a close affinity to the bacterioplankton from this mesocosm experiment, especially *Cyanobacteria* from the *Pseudanabaenaceae* and *Nostocaceae* families and *Alphaproteobacteria* from the *Rhizobiales* and *Rhodobacterales* orders (Pajares et al. 2013, 2015). The less extreme treatments (control and UV minimum tanks) had a composition and distribution of the phylogenetic groups more similar to the natural pools than to the most extreme treatments (fluctuating, 40 °C, and UV plus tanks). Bacterial mat diversity increased in the UV minimum tank treatment. Meanwhile, the extreme treatments showed drastic changes in community composition and structure, indicating a different community response to each environmental disturbance. These findings suggested that protected environments promoted complex bacterial community establishment in the microbial mats, while stressful environments reduced diversity and increased Cyanobacterial OTU dominance, mainly in *Leptolyngbya*, through environmental filtering.

In summary, the different mesocosm experiments allowed us to manipulate and replicate the CCB pools under different environmental scenarios in order to understand the sensitivity of bacterioplankton and microbial mat communities to environmental disturbances, especially fluctuating temperature. Global environmental models predict that temperatures will change sharply and in unpredictable ways, especially in deserts (see Chap. 3 in this book). Hence, the extremely ancient oasis of Cuatro Ciénegas will possibly be even frailer toward environmental change than previously expected.

What We Can Learn from What We Lost?

In 2014, "cosms" were set up at a research center in Irapuato, Mexico (Cinvestav), to study the fate of microorganisms in a small setting with periodic perturbations. While the experiments are still ongoing, in the process we have already learned more about the biology of *Bacillus* (unpublished data).

The Churince system has experienced different environmental ups and downs and is currently in a down phase, with its large desiccation lagoon being completely dry as mentioned above. Our sampling and study site in another lagoon of the Churince system (Laguna Intermedia, also known as Laguna Los Gueros) for 12 years has contained very little water. The marking poles used for the sampling sites are now all out of the water, and new poles are set up to mark the new water limits that are also dry currently lack water. Our visit to the Churince in 2016 to collect sediment and water samples in order to set up microcosms in the laboratory seemed timely, since we did not know if we would have this opportunity a few years later.

We have now three tanks at Cinvestav Irapuato where sediment has been recovering a layered structure, similar to that observed at the collection site. While it is exciting to reproduce some of the conditions of the Churince water system, it would be unfortunate if this becomes the only remnant of another lagoon in the Churince system, seeming like a white rhino in a zoo, as it might be the only ecosystem of this kind left to learn about what we lost. This situation creates a huge, not to mention sad, responsibility that no researcher wants to experience.

Conclusions

From our studies, it is clear the "cosm" experimental approaches are very powerful tools to study and evaluate different ecological factors and conditions.

The results of our various "cosm" experiments of microbes from different environments of Cuatro Ciénegas have unraveled mysteries regarding ecological variables important for microbial community resilience and the biology and specific functions of particular bacterial groups. Our results suggest that perturbations influencing community structure and community cohesion were stronger in less fluctuating environment. Our studies also serve as a warning about the impact that environmental changes caused by human activity can have on "the invisible life."

Microbial communities still carry amazing secrets about the ecology, evolution, and history of our planet, many of them of potential biotechnological value, and we hope to be able to learn more about them with our research at Cuatro Ciénegas for many years to come.

References

Adams HE et al (2010) Temperature controls on aquatic bacterial production and community dynamics in arctic lakes and streams. Environ Microbiol 12:1319–1333

Allison SD, Martiny JB (2008) Resistance, resilience, and redundancy in microbial communities. Proc Natl Acad Sci 105:11512–11519

Arroyo AI et al (2015) Plant–plant interactions as a mechanism structuring plant diversity in a Mediterranean semi-arid ecosystem. Ecol Evol 5(22):5305–5317

Baho DL et al (2012) Resistance and resilience of microbial communities–temporal and spatial insurance against perturbations. Environ Microbiol 14(9):2283–2292

Begon M et al (2006) Ecology: from individuals to ecosystems, 4th edn. Wiley, Malden

Belnap J et al (2001) Influence of biological soil crusts on soil environments and vascular plants. In: Biological soil crusts: structure, function, and management. Springer, Berlin, pp 281–300

Belnap J et al (2004) Response of desert biological soil crusts to alterations in precipitation frequency. Oecologia 141(2):306–316

Bloesch J (1988) Mesocosm studies. Hydrobiologia 159(3):221–222

Bonilla-Rosso G et al (2012) Comparative metagenomics of two microbial mats at cuatro ciénegas basin II: community structure and composition in oligotrophic environments. Astrobiology 12:659–673

Bowen JL et al (2011) Microbial community composition in sediments resists perturbation by nutrient enrichment. ISME J 5(9):1540

Boyle TP, Fairchild JF (1997) The role of mesocosm studies in ecological risk analysis. Ecol Appl 7(4):1099–1102

Breitbart M et al (2009) Metagenomic and stable isotopic analyses of modern freshwater microbialites in Cuatro Ciénegas, Mexico. Environ Microbiol 11:16–34

Cleveland CC, Liptzin D (2007) C:N:P stoichiometry in soil: is there a 'Redfield ratio' for the microbial biomass? Biogeochemistry 85:235–252

Collins SL et al (2008) Pulse dynamics and microbial processes in aridland ecosystems. J Ecol 96:413–420

Comte J, Del Giorgio PA (2010) Linking the patterns of change in composition and function in bacterioplankton successions along environmental gradients. Ecology 91(5):1466–1476

Daufresne M et al (2009) Global warming benefits the small in aquatic ecosystems. Proc Natl Acad Sci 106:12788–12793

Draggan S, Reisa JJ (1980) Controlling toxic substances: Historical perspective and future research needs. In: J. Microcosms in ecological research, Giesy P (ed). U.S. Dept. of Energy Symp. Series 52 (CONF-781101), pp. iii-xii

Dunbar J et al (1999) Levels of bacterial community diversity in four arid soils compared by cultivation and 16S rRNA gene cloning. Appl Environ Microbiol 65:1662–1669

Eichorst SA et al (2007) Isolation and characterization of soil bacteria that define Terriglobus gen. nov., in the phylum Acidobacteria. Appl Environ Microbiol 73:2708–2717

Eldridge D (2000) Ecology and management of biological soil crusts: recent developments and future challenges. Bryologist 103(4):742–747

Elser JJ et al (2005) Effects of phosphorus enrichment and grazing snails on modern stromatolitic microbial communities. Freshw Biol 50:1808–1825

Elser JJ et al (2006) Early Cambrian food webs on a trophic knife-edge? A hypothesis and preliminary data from a modern stromatolite-based ecosystem. Ecol Lett 9:295–303

Falkowski PG et al (2008) The microbial engines that drive Earth's biogeochemical cycles. Science 320(80):1034–1039

Fierer N et al (2007) Toward an ecological classification of soil bacteria. Ecology 88:1354–1364

Fierer N et al (2010) Changes through time: integrating microorganisms into the study of succession. Res Microbiol 161:635–642

Fuhrman J (2009) Microbial community structure and its functional implications. Nature 459:193–199

Garcia-Pichel F et al (2013) Temperature drives the continental-scale distribution of key microbes in topsoil communities. Science 340(6140):1574–1577

Grice GD, Reeve MR (1982a) The term mesocosm was proposed for middle-sized worlds falling between laboratory microcosms and the large, complex, real world macrocosms that includes more biological complexity. Springer, Berlin

Grice GD, Reeve MR (1982b) Introduction and description of experimental ecosystems. In: Marine mesocosms. Springer, New York, pp 1–9

Grimm NB et al (2013) The impacts of climate change on ecosystem structure and function. Front Ecol Environ 11:474–482

Gundlapally SR, Garcia-Pichel F (2006) The community and phylogenetic diversity of biological soil crusts in the Colorado Plateau studied by molecular fingerprinting and intensive cultivation. Microb Ecol 52:345–357

Häder D et al (2011) Effects of UV radiation on aquatic ecosystems and interactions with climate change. Photochem Photobiol Sci 10:242–260

Intergovernmental Panel on Climate Change (2015) Climate change 2014: mitigation of climate change, vol 3. Cambridge University Press, Cambridge

Janssen PH (2006) Identifying the dominant soil bacterial taxa in libraries of 16S rRNA and 16S rRNA genes. Appl Environ Microbiol 72:1719–1728

Jessup CM et al (2004) Big questions, small worlds: microbial model systems in ecology. Trends Ecol Evol 19:189–197

Jessup CM et al (2005) Microbial experimental systems in ecology. Adv Ecol 37:273–307

Johnson SL et al (2012) Increased temperature and altered summer precipitation have differential effects on biological soil crusts in a dryland ecosystem. Glob Chang Biol 18(8):2583–2593

Kuske CR et al (1997) Diverse uncultivated bacterial groups from soils of the arid southwestern United States that are present in many geographic regions. Appl Environ Microbiol 63:3614

Langenheder S, Szekely AJ (2011) Species sorting and neutral processes are both important during the initial assembly of bacterial communities. ISME J 5:1086–1094

Langenheder S et al (2005) Weak coupling between community composition and functioning of aquatic bacteria. Limnol Oceanogr 50(3):957–967

Levins R (1966) The strategy of model building in population biology. Am Sci 54(4):421–431

Lindström ES, Langenheder S (2012) Local and regional factors influencing bacterial community assembly. Environ Microbiol Rep 4(1):1–9

Lopez-Lozano NE et al (2012) Bacterial communities and the nitrogen cycle in the gypsum soils of Cuatro Cienegas basin, Coahuila: a Mars analogue. Astrobiology 12:699–709

Lopez-Lozano NE et al (2013) Microbial secondary succession in soil microcosms of a desert oasis in the Cuatro Cienegas Basin, Mexico. PeerJ 1:e47

Maier S et al (2014) Analyses of dryland biological soil crusts highlight lichens as an important regulator of microbial communities. Biodivers Conserv 23(7):1735–1755

Makhalanyane TP et al (2013) Evidence of species recruitment and development of hot desert hypolithic communities. Environ Microbiol Rep 5(2):219–224

Maranger R et al (2002) Accumulation of damaged bacteria and viruses in lake water exposed to solar radiation. Aquat Microb Ecol 28:213–227

Martiny JBH et al (2006) Microbial biogeography: putting microorganisms on the map. Nat Rev Microbiol 4(2):102
Nagy ML et al (2005) The prokaryotic diversity of biological soil crusts in the Sonoran Desert (Organ Pipe Cactus National Monument, AZ). FEMS Microbiol Ecol 54:233–245
Nemergut DR et al (2013) Patterns and processes of microbial community assembly. Microbiol Mol Biol Rev 77(3):342–356
Nunes da Rocha U et al (2015) Isolation of a significant fraction of non-phototroph diversity from a desert biological soil crust. Front Microbiol 6:277
Odum EP (1983) Basic ecology. Saunders College Pub, Philadelphia
Pajares S et al (2012) Mesocosms of aquatic bacterial communities from the Cuatro Cienegas Basin (Mexico): a tool to test bacterial community response to environmental stress. Microb Ecol 64:346–358
Pajares S et al (2013) Drastic changes in aquatic bacterial populations from the Cuatro Cienegas Basin (Mexico) in response to long-term environmental stress. Antonie Van Leeuwenhoek 104:1159–1175
Pajares S et al (2015) Multivariate and phylogenetic analyses assessing the response of bacterial mat communities from an ancient oligotrophic aquatic ecosystem to different scenarios of long-term environmental disturbance. PLoS One 10:1–18
Reed HE, Martiny JB (2013) Microbial composition affects the functioning of estuarine sediments. ISME J 7(4):868
Roesch LFW et al (2007) Pyrosequencing enumerates and contrasts soil microbial diversity. ISME J 1:283–290
Santos AL et al (2012) Effects of UV-B radiation on the structural and physiological diversity of bacterioneuston and bacterioplankton. Appl Environ Microbiol 78:2066–2069
Souza V et al (2006) An endangered oasis of aquatic microbial biodiversity in the Chihuahuan desert. Proc Natl Acad Sci U S A 103:6565–6570
Souza V et al (2008) Microbial endemism: does phosphorus limitation enhance speciation? Nat Rev Microbiol 6:559–564
Souza V et al (2012) The Cuatro Ciénegas Basin in Coahuila, Mexico: an astrobiological Precambrian park. Astrobiology 12:641–647
Steven B et al (2013) Dryland soil microbial communities display spatial biogeographic patterns associated with soil depth and soil parent material. FEMS Microbiol Ecol 86(1):101–113
Stewart RI et al (2013) Mesocosm experiments as a tool for ecological climate-change research. Adv Ecol Res 48:71–181
Strickland MS et al (2009) Testing the functional significance of microbial community composition. Ecology 90:441–451
Wall DH, Virginia RA (1999) Controls on soil biodiversity: insights from extreme environments. Appl Soil Ecol 13:137–150
Whitman WB et al (1998) Perspective prokaryotes: the unseen majority. Proc Natl Acad Sci U S A 95:6578–6583
Widder S et al (2016) Challenges in microbial ecology: building predictive understanding of community function and dynamics. ISME J 10:2557–2568
Wijngaarden RPV et al (2005) Threshold levels for effects of insecticides in freshwater ecosystems: a review. Ecotoxicology 14(3):355
Wohl DL et al (2004) Functional redundancy supports biodiversity and ecosystem function in a closed and constant environment. Ecology 85(6):1534–1540
Yeager CM et al (2004) Diazotrophic community structure and function in two successional stages of biological soil crusts from the Colorado plateau and Chihuahuan desert. Appl Environ Microbiol 70:973–983
Zelikova TJ et al (2012) Warming and increased precipitation frequency on the Colorado Plateau: implications for biological soil crusts and soil processes. Plant Soil 355(1-2):265–282

Chapter 8
Single Genus Approach to Understanding Bacterial Diversity, Niche, Distribution, and Genomics: The *Bacillus* in Cuatro Ciénegas

Maria Dolores Rodriguez Torres, Luis David Alcaraz, Valeria Souza, and Gabriela Olmedo-Álvarez

Abstract To understand microbial life and get the full picture, we ought to zoom out in order to grasp its magnitud and diversity and zoom in to dissect their genes and functions. We have obtained a significant amount of data on the great diversity of bacteria in the different environments of the Cuatro Ciénegas Basin (CCB), equivalent to obtaining the census of a city. However, what are these bacteria doing and what is their role? How do they get along? We have chosen the genus *Bacillus* to answer these questions. *Bacillus* can form highly resistant spores that disperse readily and can remain dormant for years, which complicates their study. On the other hand, bacteria of the *Bacillus* genus are abundant, diverse, and easily cultivated, so they are a good model, both to understand their biology and to study their interactions in communities. With 2500 isolates from CCB and data obtained from microbiological and genomic approaches, we can now write *Bacillus* tale in Cuatrociénegas.

M. D. R. Torres · G. Olmedo-Álvarez (✉)
Laboratorio de Biología Molecular y Ecología Microbiana, Departamento de Ingeniería Genética, Unidad Irapuato, Centro de Investigación y de Estudios Avanzados IPN, Irapuato, Mexico
e-mail: golmedo@ira.cinvestav.mx

L. D. Alcaraz
Departamento de Biología, Facultad de Ciencias, Universidad Nacional Autónoma de México, México City, Mexico

V. Souza
Laboratorio de Evolución Molecular y Experimental, Departamento de Ecología Evolutiva, Instituto de Ecología, Universidad Nacional Autónoma de México, Mexico City, Mexico

V. Souza et al. (eds.), *Cuatro Ciénegas Ecology, Natural History and Microbiology*, Cuatro Ciénegas Basin: An Endangered Hyperdiverse Oasis,
https://doi.org/10.1007/978-3-319-93423-5_8

Exploring Conservation of Lineage Phenotypic Traits Within the *Bacillus* Genus in a Community Setting

Bacteria of the *Bacillus* genus have been extensively characterized with classical microbiology, biochemistry, genomic, and proteomic approaches. The *Bacillus* species are considered generalists, because they are present in very diverse environments, have versatile metabolic capabilities for exploiting different organic and inorganic substrates, and have excellent dispersion capacity due to their ability to form spores. Spores are quiescent and highly resistant structures able to withstand nutritional and environmental stress that allows them to thrive in different environments. Although *Bacillus subtilis* is the model species, studies have been done in many species within the genus involving sporulation, and other traits have been observed to be deeply rooted among the *Firmicutes*. These traits include Gram-positive staining as well as possessing capabilities for swimming, swarming, and biofilm formation.

Bacteria in the *Bacillus* genus are abundant, diverse, and easily cultivated. From metagenomic studies at the Churince water system, we have (unpublished data), we have found that the *Bacillus* species represent 2.4% of the microbial abundance (unpublished data). *Bacillus* species from CCB do not form a monophyletic group and exhibit ancient divergences, this can be said due to the isolation of lineages that could have inhabited an ancient ocean (Moreno-Letelier et al. 2012). Souza et al. (2006) found that nearly 50% of phylotypes from the CCB were closely related to organisms from marine environments, including some *Bacillus* species. For some of the isolated *Bacillus* strains from Cuatro Cienegas, their closest relatives are from the Gulf of Mexico (*Bacillus* sp. SG-1, *Bacillus* sp. NRRL B-14911), Sea of Korea (*B. aquimaris*), and tidal flat in the Yellow Sea (*B. marisflavi*) or isolated from marine organisms (*B. horikoshii*).

It has also been noted that there are several endemic *Bacillus* species within the Cuatro Cienegas Basin, most of them with very old divergence (Moreno-Letelier et al. 2012). Although we know that many Bacillus lineages co-occur in this environment (Fig. 8.1), we still have many questions: What are these bacteria doing? What is their role? How do they get along? Is taxonomical diversity the same as functional diversity?

The results from this first sequenced genome CCB was *Bacillus coahuilensis* m4-4 (Alcaraz et al. 2008), which was a strain isolated from the water column of a desiccation lagoon that no longer exists (Laguna Grande) in the Churince by Cerritos et al. (2008). The sequence revealed that this bacterium had many unexpected features that had never before been observed in the *Bacillus* genus. It lacked teichoic acid, a phosphate-rich cell wall layer always present in Gram-positive bacteria. It also possessed a bacteriorhodopsin gene, which had never been described for *Bacillus*, and, notably, it possessed genes for the biosynthesis of sulfolipids (Alcaraz et al. 2008). These genomic features allowed *B. coahuilensis* to add sulfolipids to its membrane, and, since it had no phosphorus-rich teichoic acid requirement, these bacteria can cope with the limiting phosphorus conditions characteristic

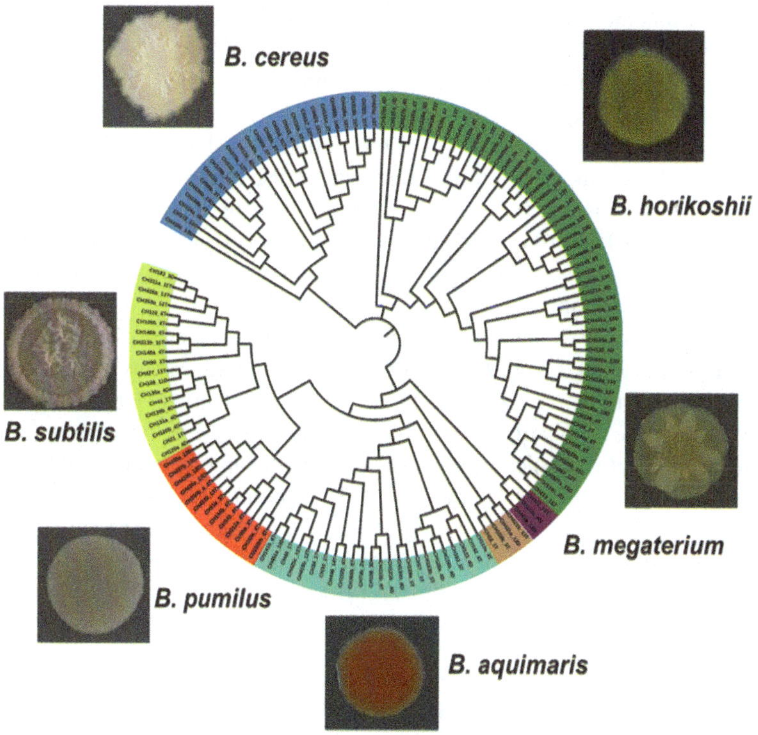

Fig. 8.1 Phylogenetic reconstruction of 465 thermo-resistant isolates from the Churince aquatic system. Most of the isolates are related to *Bacillus* species

of CCB. The results from this first genome were very interesting, as they showed us that the genomic data reflected the metabolic potential of the bacteria and the ecology of the site.

Bacillus m3-13 was the second CCB strain to be sequenced, also from the same extinct lagoon from which *B. coahuilensis* was isolated, and with it, we began the comparative exploration of aquatic CCB *Bacillus*. Before our work, most *Bacillus* studies had been done on the so-called soil species, namely, *B. cereus* and *B. subtilis/pumilus*. We were able to do the first phylogenomic analysis of an aquatic *Bacillus* genus and describe the first core genome for the genus consisting of 870 genes (Alcaraz et al. 2010). We showed that even for the development of the spore—a highly conserved process—the most conserved genes were those driving transcription at different stages, while the signaling and structural genes showed considerable variability (Alcaraz et al. 2010).

The relationship between the phylogeny and the functional features of each member of the *Bacillus* group revealed great variability that limited the predictive power of a microbial composition approach (Alcaraz et al. 2010; Martiny et al. 2013). According to Philippot et al. (2010), the consistency of traits can be maintained up to the phylum level. Nevertheless, other reports support the idea that the ecological traits may be specific to species or strains, since the microbial genomes are highly dynamic and can change rapidly through gene loss, gene duplication, and acquisition of genes from distant lineages via horizontal gene transfer (HGT, Boon et al. 2014). It has been shown that strains within a given species may have between 30 and 35% gene content differences (Konstantinidis et al. 2006). Bacterial species are known to have a mosaic of transferred genes since their high genetic diversity is obtained in some cases from distantly related organisms (Ochman et al. 2000). If there are at least 1000 different genes at the species level, numerous phenotypic differences can be expected. With these differences, it is not surprising to find species colonizing different habitats.

Most comparative studies on bacterial traits and genomes are based on strains that have been sampled from highly diverse environments. Many functional traits have been shown to be phylogenetically dispersed in bacteria (Martiny et al. 2013). These studies have shown that lifestyle differences correlate with variations in genes required to interact with specific environments. For example, features of free-living bacteria differ from those of pathogens and symbionts. On the other hand, for a community to remain stable, it is suggested that the synthesis of products found in the environment may be sufficient to support a member's loss of genes or functions (McInerney et al. 2011; Morris et al. 2012; Boon et al. 2014).

Adaptations to Deal with Very Low Concentrations of Phosphorus in CCB

Phosphorus (P) is an essential element for multiple processes and is found in different essential molecules, such as DNA, cell membranes, and cell walls (teichoic acid in Gram-positive bacteria), and in molecules for energy transfer. Different *Bacillus* taxonomic groups coexist in the Churince water system of CCB, a site with high concentrations of sulfates, magnesium, and calcium but low concentrations of NaCl and carbonates and, notably, extremely low concentrations of phosphorus (0.3 μM). Therefore, this water system is categorized as an ultra-oligotrophic environment (Cerritos et al. 2008).

Tapia-Torres et al. (2016) investigated how bacteria in CCB managed to obtain P. They measured the availability of P and the enzymatic activity associated with P release in soil and sediment from the Churince and Pozas Azules systems. They also evaluated the use of different sources of phosphorus for culturable bacteria. The patterns of enzymatic expression showed substantial differences in the amounts and types of exoenzymes in the different microbial communities in the study sites. Phosphodiesterase activity was higher in soil samples, phosphomonoesterase activity was higher in sediment, and phosphonatase activity was detected in soil samples but not in sediment samples.

Regarding the use of different sources of phosphorus for culturable bacteria, such as DNA, calcium phosphate, phosphite, and phosphonates (2PA or 2AEP), Tapia-Torres et al. (2016) results showed no correlation between taxonomic group and the ability to use different P sources. This lack of correlation suggests that the genes involved in the utilization of P can transfer laterally among strains. This was corroborated by our studies on the gene encoding the Pst transporter, that suggests that it moves among bacteria through lateral transfer (Moreno-Letelier et al. 2011).

Since the different *Bacillus* species can break down and use P forms in different oxidation states, they contribute to P cycling in the ecosystem, suggesting a dynamic movement of P utilization traits among bacteria in microbial communities. Most bacteria can use potassium phosphate, calcium phosphate, and DNA and to a lesser extent phosphite, 2PA, or 2AEP. Interestingly, members of the *B. cereus* group isolated from aquatic sediments at the Churince could use more phosphorus sources compared to members isolated from soils. The ability to use phosphite was high in members isolated from sediment. This phosphite utilization preference among *B. cereus* from sediment reflects the variable occurrence of the different P sources in soil and sediment and shows evidence for the presence and importance of phosphite in natural environments (Tapia-Torres and Olmedo 2018). On the other hand, the dispersion of traits related to P acquisition within and across taxonomic groups is suggestive of niche partitioning strategies that also explain how genetic diversity is maintained.

Other physiological adaptations have been observed in *Bacillus* species that live in the phosphorus-limited CCB environment. The best characterized strains are those of *B. coahuilensis*, which were isolated from the water column of a shallow desiccation lagoon in the Churince system (m4-4 and m2-6) and Pozas Azules (p1.1.43) (see Chap. 5 in this book). These strains possess genes that encode key enzymes in the synthesis of sulfolipids (sulfoquinovose synthase and glycosyltransferase). Sulfolipids can be integrated into the membrane and can replace some phospholipids. *B. coahuilensis* also lacks genes involved in the synthesis of teichoic acid, a polymer located within the Gram-positive cell wall, and involved in the regulation of cell division and Gram-positive bacterial physiology (Gómez-Lunar et al. 2016; Brown et al. 2013). This finding helped to reveal that some *Bacillus* taxonomic groups related to marine bacteria environments and isolated from CCB were Gram-negative (unpublished data).

Although *B. coahuilensis* m4-4 and m2-6 strains share the same environment as *Bacillus* m3-13, *Bacillus* m3-13 shares many strategies with the m2-6 strain for dealing with the poor nutrient environment, but not the m4-4 strain. *Bacillus* m3-13 encodes phosphonate ABC importers, permeases, and phosphonate lyases that can take up and assimilate phosphonates (see Chap. 5 in this book). In this sense, the *Bacillus* species contain adaptations that reveal different strategies for using phosphorus in order to survive in the oligotrophic CCB environment.

Traits Deeply Rooted in Taxonomic Lineages: Exploring Conservation of Lineage Phenotypic Traits Within a *Bacillus* Genus in a Community Setting

One of the most important characteristics of the *Bacillus* species when facing nutrients scarcity is their ability to form endospores: structures highly resistant to heat, radiation, desiccation, and oxidizing agents. Most of the *Bacillus* species obtained from Cuatro Cienegas were isolated by heating environmental samples, so it was assumed that the strains readily sporulated. However, when a representative group of different taxonomic groups were subjected to different culture media to trigger sporulation, the strains related to aquatic members (i.e., *B. marisflavi*, *B. aquimaris*, and *B. horikoshii*) had sporulation values below 1% (unpublished data), whereas *Bacillus* related to *B. subtilis*, *B. thuringiensis*, or *B. pumilus* had a sporulation frequency that was above 50% and could reach 100% (unpublished data). Based on genomic data, Alcaraz et al. (2010) observed that less than one-third of the genes involved in the sporulation of *B. subtilis* were conserved across other *Bacillus* species with essential differences in the signaling genes.

Genomic and phenotypic insights of *B. coahuilensis* showed that its genome is a genetic mosaic. It has acquired genes involved in phosphorus utilization efficiency and genes relevant to adapting to high-light environments possibly through HGT from cyanobacteria (Alcaraz et al. 2008). These traits depend on a few genes, like other traits related to nutrition that are known to be easily transferred among bacteria (Martiny et al. 2013). In this sense, analysis of the utilization of carbon sources in *B. coahuilensis* strains showed that m2-6 and p1.1.43 had similar utilization patterns of poly- and oligosaccharides, carboxylic acids, and an aromatic compound, but m4-4 and m2-6 were more similar in their choice of monosaccharide utilization. In general, *Bacillus* species exhibit some differences in their preference for different carbon sources, according to genomic predictions made from the presence of transporters in genomes belonging to (1) soil members (*B. subtilis*/*B. licheniformis* and *B. pumilus*), (2) aquatic members (*B. marisflavi*, *B. aquimaris*, and *B. horikoshii*), and (3) *B. cereus*-related species (Alcaraz et al. 2010). However, experimental evaluation of members from these groups showed diverse capabilities to utilize glucose, raffinose, trehalose, xylose, and sorbitol. These capabilities were sparsely distributed and did not cluster with any particular *Bacillus* lineage (Rodriguez-Torres et al. 2017).

The *Bacillus* species possess other essential traits associated with specific growth or developmental stages. Swarming and swimming are traits related to motility (Fig. 8.2). Both traits depend on a multi-protein complex that forms the flagella. Swimming in an aqueous medium allows an individual bacterium to move toward chemoattractants or to get free from a biofilm. Swarming allows bacterial groups to move on surfaces with less water, so this trait is tested on semisolid agar. Swarming requires additional flagella as well as surfactants to break the superficial tension (Kearns and Losick 2003). The ability to make biofilms is another conserved trait in bacteria. Biofilms allow bacteria to attach to surfaces and to grow as a community so as to produce a barrier against environmental hazards (i.e., antibiotics). Biofilm production depends on multiple genes considered deeply rooted in taxonomic lineages, such as exopolysaccharides. Swarming and biofilm formation are multicellular traits. These traits are considered social features (González-Pastor et al. 2003), since they provide benefits at a group level and depend on the coordinated expression of genes and signals (Stoodley et al. 2002; Daniels et al. 2004). Swimming and sporulation may also respond to group signals.

The results of the swarming tests of the *Bacillus* isolates from Churince were negative for aquatic members, but positive for soil members, while a smaller number of *B. cereus*-related species presented this ability (Rodriguez-Torres et al. 2017). The three *B. coahuilensis* strains lacked swarming capabilities, and their genomic analysis showed that they lacked the *srfABCD* operon that encodes surfactin (Gómez-Lunar et al. 2016). Regarding swimming evaluation—depending only on flagella synthesis—most species showed this ability, except for *B. coahuilensis* m4-4 and 9 out of 23 strains related to the *B. horikoshii* (Fig. 8.2).

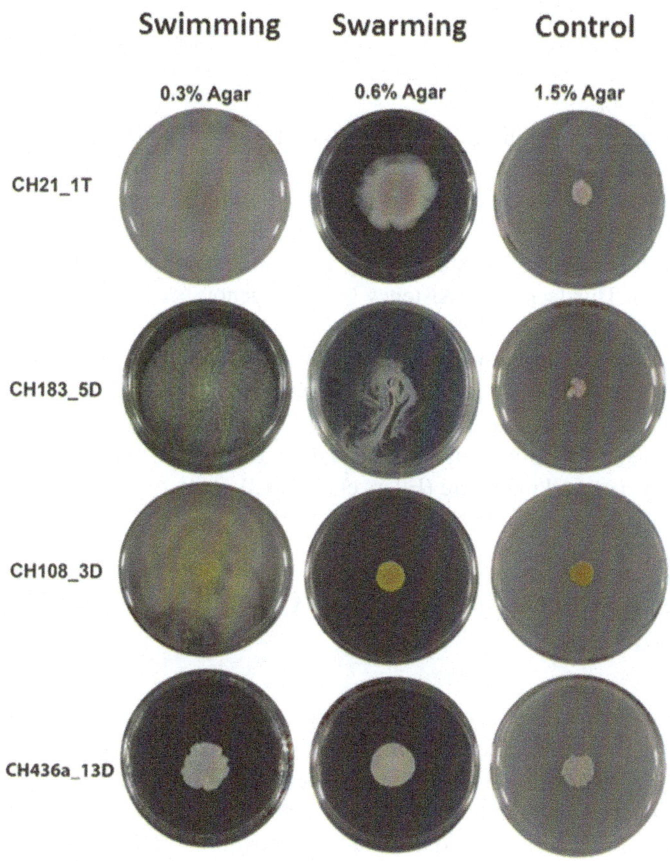

Fig. 8.2 Swarming and swimming motility evaluated in *Bacillus* strains from the Churince. The plates show photographs of strains that exhibit swarming and swimming motility compared with normal agar (1.5%: first column), the second column shows plates with 0.6% agar (to evaluate swarming), and the third column shows with 0.3% agar (to evaluate swimming). We show three types of results: (1) strain CH21_1T y CH183_5D exhibited swarming and swimming, (2) strain CH108_3D exhibited swimming but not swarming, and (3) strain CH436a_13D did not present either swarming or swimming motility

Biofilm formations are organized populations that depend on the expression of an extracellular polymeric substance (Spoering and Gilmore 2006). In natural environments, biofilm formations typically provide benefits at the group level (Stoodley et al. 2002; Daniels et al. 2004). Biofilm results of *Bacillus* isolated from CCB showed results similar to swarming, since it was usually present in members of the soil *Bacillus* and the *B. cereus* group but absent in most aquatic members (Fig. 8.2).

Bacteria also possess metabolic routes for producing their amino acids. Isolates that can produce all amino acids are called prototrophs, while those that cannot produce one or more amino acids are called auxotrophs. The evaluation of amino acid biosynthesis capabilities revealed this to be another variable trait among members of the communities. There were numerous auxotrophs found in *B. cereus* and also among aquatic *Bacillus*. In contrast, *B. subtilis/pumilus* members were almost always prototrophs.

We analyzed the genomes of some bacteria to further understand the nature of auxotrophs that lacked particular genes in the amino acid biosynthetic pathways. For *B. coahuilensis*, some auxotrophs correlated with the absence of genes in the conventional routes (Gómez-Lunar et al. 2016). The biosynthesis routes appeared to be complete for *B. cereus*, but surprisingly some auxotrophs were observed (unpublished results). This lack of consistency between genotypes and phenotypes has to be further explored.

How can this trait variation in *Bacillus* from the communities be explained? The variation observed could be attributed to compensated trait loss occurring in communities which allows an individual, even with a function loss, to continue obtaining the trait from the community or an adaptive trait that responds to specific cues that cannot be assayed in culture (Ellers et al. 2012).

Antagonistic Interactions Influence Community Assemblage at Each Site

As explained above, the absence of some traits among strains of the sediment communities suggests that particular product loss can be afforded if these can be obtained by sharing with other community members. However, sharing it is not the only thing happening in communities. It has been suggested that the interactions among culturable bacteria are dominated by competition (Foster and Bell 2012).

Pérez-Gutiérrez et al. (2013) analyzed the interactions between several coexisting lineages of *Bacillus* species from five physicochemically similar sampling sites from the Churince system. They determined the antagonism potential of the isolates and compared them against each other, as well as within and across the different sampling sites to understand the assembly of sediment communities and the influence of interfering interactions. A directed network could describe the 6000 scored interactions with a hierarchical structure that revealed a food chain-like structure in which different *Bacillus* taxonomic groups occupied specific positions. Fewer

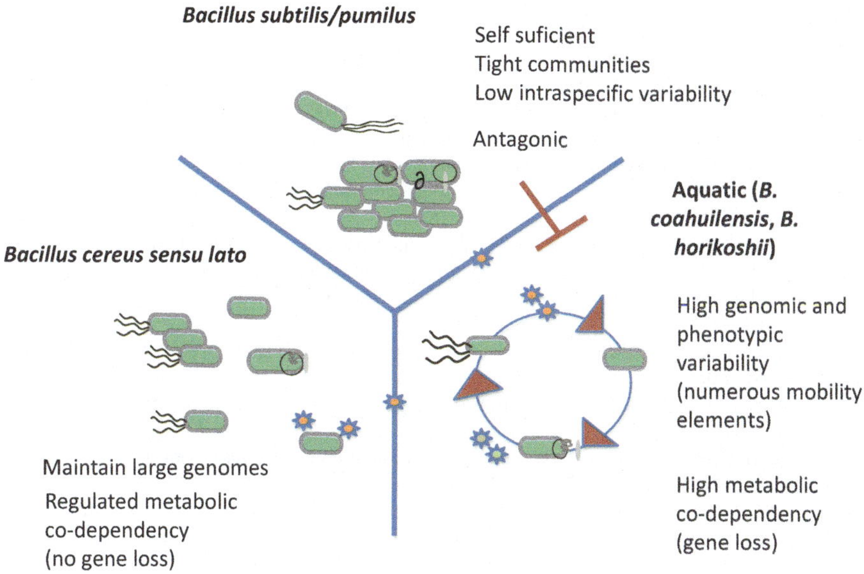

Fig. 8.3 Distinct traits associated to the different *Bacillus* taxonomic groups from sediment communities. Three main strategies were observed among the *Bacillus* taxonomic groups studied. The *Bacillus subtilis/pumilus* exhibits low intraspecific variability and are also the most antagonistic. The aquatic bacteria are generally sensitive to antagonism and exhibit great variability in social traits. The group of the *B. cereus* is generally resistant to antagonism, maintains large genomes, but exhibits dependency on the communities to complement nutritional and social traits

antagonistic interactions were detected within sites than across sites, suggesting that antagonistic interactions influenced community assemblage at each site. Furthermore, antagonism and sensitivity could be associated with specific *Bacillus* taxonomic groups, and the data could be represented as a hierarchical network representation. For instance, the *B. subtilis/pumilus* group was usually antagonistic toward the aquatic *Bacillus aquimaris/horikoshii*, while the *B. cereus* group members were resistant to antagonism and were not antagonistic (Fig. 8.3).

In summary, our in-depth analysis of a single genus, the *Bacillus* from CCB, allowed us to unravel wide intraspecific variability among community members at the genomic and phenotypic level focusing on substrate utilization traits and social traits (Fig. 8.3). We also showed that the ecological strategies are different at the species level. Although there is competition, different members diversify their micro-niche allowing them to coexist in complex communities. New studies need to elucidate how those complex communities get build, which is the role of historic vs change in the community assemblage, not only of the coexisting *Bacillus* but of all the community. This is a dauntingly fascinating challenge.

Acknowledgments This chapter was written during a sabbatical leave of VS in the Department of Ecology, Evolution, and Behavior, University of Minnesota, in Dr. Michael Travisano's laboratory, supported by the program PASPA-DGAPA, UNAM.

References

Alcaraz LD et al (2008) The genome of *Bacillus* coahuilensis reveals adaptations essential for survival in the relic of an ancient marine environment. Proc Natl Acad Sci 105(15):5803–5808

Alcaraz LD et al (2010) Understanding the evolutionary relationships and major traits of *Bacillus* through comparative genomics. BMC Genomics 26:311–332

Boon E et al (2014) Interactions in the microbiome: communities of organisms and communities of genes. FEMS Microbiol Rev 38(1):90–118

Brown S et al (2013) Wall teichoic acids of gram-positive bacteria. Annu Rev Microbiol 67:313–336

Cerritos R et al (2008) *Bacillus coahuilensis* sp. nov., a moderately halophilic species from a desiccation lagoon in the Cuatro Ciénegas Valley in Coahuila, Mexico. Int J Syst Evol Microbiol 58(4):919–923

Daniels R et al (2004) Quorum sensing and swarming migration in bacteria. FEMS Microbiol Rev 28(3):261–289

Ellers J et al (2012) Ecological interactions drive evolutionary loss of traits. Ecol Lett 15(10): 1071–1082

Foster KR, Bell T (2012) Competition, not cooperation, dominates interactions among culturable microbial species. Curr Biol 22(19):1845–1850

Gómez-Lunar Z et al (2016) Microevolution analysis of *Bacillus coahuilensis* unveils differences in phosphorus acquisition strategies and their regulation. Front Microbiol 7:58

González-Pastor JE et al (2003) Cannibalism by sporulating bacteria. Science 301(5632):510–513

Kearns BD, Losick R (2003) Swarming motility in undomesticated *Bacillus subtilis*. Mol Microbiol 49(3):581–590

Konstantinidis KT et al (2006) The bacterial species definition in the genomic era. Philos Trans R Soc Lond Ser B Biol Sci 361(1475):1929–1940

Martiny AC et al (2013) Phylogenetic conservatism of functional traits in microorganisms. ISME J 7(4):830–838

McInerney JO et al (2011) The public goods hypothesis for the evolution of life on Earth. Biol Direct 6:41

Moreno-Letelier A et al (2011) Parallel evolution and horizontal gene transfer of the pst operon in firmicutes from oligotrophic environments. Int J Evol Biol 2011:10

Moreno-Letelier A et al (2012) Divergence and phylogeny of firmicutes from the Cuatro Ciénegas Basin, Mexico: a window to an ancient ocean. Astrobiology 12(7):674–684

Morris JJ et al (2012) The black queen hypothesis: evolution of dependencies through adaptive gene loss. MBio 3(2):e00036-12

Ochman H et al (2000) Lateral gene transfer and the nature of bacterial innovation. Nature 405(6784):299–304

Pérez-Gutiérrez RA et al (2013) Antagonism influences assembly of a *Bacillus* guild in a local community and is depicted as a food-chain network. ISME J 7(3):487–497

Philippot L et al (2010) The ecological coherence of high bacterial taxonomic ranks. Nat Rev Microbiol 8:523–529

Rodriguez-Torres MD et al (2017) Phenotypic microdiversity and phylogenetic signal analysis of traits related to social interaction in *Bacillus* spp. from sediment communities. Front Microbiol 8:29

Souza V et al (2006) An endangered oasis of aquatic microbial biodiversity in the Chihuahuan desert. Proc Natl Acad Sci 103(17):6565–6570

Spoering AL, Gilmore MS (2006) Quorum sensing and DNA release in bacterial biofilms. Curr Opin Microbiol 9:133–137

Stoodley P et al (2002) Biofilms as complex differentiated communities. Annu Rev Microbiol 56:187–209

Tapia-Torres Y, Olmedo G (2018) Life on phosphite: a metagenomics tale. Trends Microbiol 26(3):170–172

Tapia-Torres Y et al (2016) How to live with phosphorus scarcity in soil and sediment: lessons from bacteria. Appl Environ Microbiol 82(15):4652–4662

Chapter 9
Bioprospecting in Cuatro Cienegas: A Set of Open Possibilities to Save the Ecosystem

Héctor Arocha-Garza, Hamlet Avilés-Arnaut, Nahui Olin Medina-Chávez, Mario Figueroa, Valeria Souza, and Susana De la Torre-Zavala

Abstract The unique and untapped microbiological diversity of the Cuatro Cienegas Basin needs to be saved from the overexploitation of the aquifer by unsustainable agricultural practices. Considering the great value of the biodiversity enclosed in this valley, as a group we decided to change the future of the area, to start evolving from a poor agricultural site, to a future biotech hub, where high school students would be the "first to open the treasure box" of the biotechnological potential. By following the Nagoya treaty, we can create value for their biodiversity and return to their community the necessary resources for education and sustainable development to alleviate poverty. But before setting students into the search, we needed to see if the dream was possible. We all knew that *Actinobacteria* were the obvious first suspect due to the large diversity of secondary metabolites that this lineage of bacteria has been reported to produce. Through our research, we have found an enormous "hidden potential," solutions that range from new antibiotics, antifungals, cancer antagonists, energy potential by degrading cellulose, and sustainable agriculture possibilities with probiotics plus a series of untapped possibilities in methanogens and halophilic Archaea. This is just the beginning of a children-driven revolution. We hope that this potential is channeled to a better society, to a university in the local town where the biotech potential is directly explored and exploited by the owners of the land.

H. Arocha-Garza · H. Avilés-Arnaut · N. O. Medina-Chávez · S. De la Torre-Zavala (✉)
Facultad de Ciencias Biológicas, Instituto de Biotecnología, Universidad Autónoma de Nuevo León, San Nicolás de los Garza, Nuevo León, Mexico
e-mail: susana.delatorrezv@uanl.edu.mx

M. Figueroa
Departamento de Farmacia, Facultad de Química, Universidad Nacional Autónoma de México, Ciudad de México, Mexico

V. Souza
Departamento de Ecología Evolutiva, Instituto de Ecología, Universidad Nacional Autónoma de México, Coyoacán, Mexico

V. Souza et al. (eds.), *Cuatro Ciénegas Ecology, Natural History and Microbiology*, Cuatro Ciénegas Basin: An Endangered Hyperdiverse Oasis,
https://doi.org/10.1007/978-3-319-93423-5_9

Why Is Cuatro Cienegas Basin a Perfect Bioprospecting Site?

As previously described, Cuatro Cienegas Basin (CCB) is as a hot spot of singular micro-biodiversity which traits and characteristics reveal mechanisms of adaptation and organization that have allowed the survival of stromatolites, microbial mats, and microbial communities with ancestral marine affinities (Souza et al. 2006; Alcaraz et al. 2008; Moreno-Letelier et al. 2012) (see Chap. 1 in this book). This survival is the testimony of their ability to successfully thrive and evolve in a harsh, nutrient-depleted, isolated environment for hundreds of millions of years.

Such unique, untapped microbiological diversity suddenly became almost a provocation that made our research group come up to CCB with questions regarding the possible biotechnological applications of this "lost world." We all knew that *Actinobacteria* was the obvious "pick" for searching natural active compounds, but the challenge was enormous, and we started planning different strategies. The research group led by Dr. Francisco Barona from Unidad de Genómica Avanzada, CINVESTAV, started the first effort on bioprospecting actinobacteria from CCB using genome mining and phylogenomics as discovery tools (Cruz-Morales et al. 2017).

Our group, working at the Universidad Autónoma de Nuevo León (Team Nuevo Leon) in collaboration with researchers at UNAM, tackled the challenge of bioprospecting actinobacteria taking the culturable "traditional" approach. However, we wanted to avoid redundancy when isolating actinomycetes and searching for bioactive compounds, so we targeted actinomycetes, designing a selective isolation strategy based on the use of a wide array of culture media to retrieve diverse but also "rare" and "gifted microbes" (Baltz 2016; Tiwari and Gupta 2012) from the CCB ponds. Before starting to look for *Actinobacteria* that produced bioactive compounds, we carried out a biodiversity assessment.

Our approach included the isolation and analysis of both environmental DNA and 16S rRNA. Through these simultaneous studies, we showed that the *Actinobacteria* were among the 50 most abundant phylotypes in Churince, while, for example, the *Bacillus* did not make it to that list. The *Actinobacteria* were present in both water and sediments, and there was no change in frequency in response to a nutrient input, while many proteobacteria, in particular endemic ones, disappeared after nutrient addition (Lee et al. 2015).

As we started culturing bacteria and sequencing the first 16S rRNA tags, we soon realized that the *Actinobacteria* from CCB were as unique as many of the other cultivated taxa previously described in this area (Escalante et al. 2009; Cerritos et al. 2011; Rebollar et al. 2012; Avitia et al. 2014; Toribio et al. 2011; Dominguez-Escobar et al. 2011). We later confirmed our observations, identifying lineages of *Streptomyces* isolated from CCB that formed six monophyletic clades in a deeply represented and well-supported phylogeny of the family *Streptomycetaceae*, clearly suggesting endemicity (Arocha-Garza et al. 2017).

Genomic plasticity through mutations, horizontal gene transfer, and transposons result in divergence and speciation. The evident high endemicity makes CCB microorganisms' secondary metabolites unique and different, hence attractive from the biotechnological point of view.

In this chapter, we present an overview of the biotechnological potential of actinomycetes and other microbes existing in CCB. Curiously, we are not driven by the usual greed that is associated with bioprospecting, but by the belief that this can become the most powerful tool for its conservation. Ecosystem services may be a nice term generally used to justify protecting any beautiful place on Earth, but here we had a true *lost world*, full of treasures that were so far intangible.

Moreover, we started our bioprospecting project by following the terms established by the Nagoya treaty (2010), an international treaty that seeks that, before any bioprospection, the community and owners of the land know and acknowledge the biotechnological development. After that first step, both scientist and potential commercial ventures also acknowledge that if any genetic resources result in commercial benefits, those will be shared with the community for sustainable development (ABS, access and benefit sharing).

Our efforts in bioprospecting in the CCB area, came up as a desire to create value for this biodiversity that must be preserved but also a genuine need to continue pushing the border of knowledge in this still vastly unexplored area in Mexico. Trailblazing this particularly political "conflicting" area, we are expecting more than anything else that this effort will be positive for both the community of people within CCB and the conservation of the wetland.

The High schoolers from CBTA No. 22 are our focus group (Centro Bachillerato Técnico Agropecuario), these kids that are an integral part of this bioprospecting effort and their awareness of the richness of this amazing place makes us think that conservation of this oasis is possible. They are the stewards of the future (see Chap. 1 in this book, in particular Fig. 1.1 t-w).

Actinobacteria: The First Obvious Suspect

Among the microbial world, the phylum *Actinobacteria* and particularly *Actinomycetes* diversity play a major ecological role in the oligotrophic environment of CCB. Historically, the *Actinomycetes* have been considered to be the most productive sources for novel natural products since, regardless the site of isolation, these produce a plethora of secondary metabolites of a wide variety of industrial and biomedical applications. In a survey of 100 of the most important metabolites produced by bacteria and fungi, 68% were produced by the *Streptomyces* species, 15% by other actinomycetes, 5% by other bacteria, 12% by fungi, and 0% by archaea (Katz and Baltz 2016).

Despite the history of success in bioprospecting actinomycetes for therapeutics in human and animal health, as well as plant-crop protection (among other industrial applications), the search for novel and relevant metabolites decreased in the last decades due to the redundancy in the natural product discovery (Katz and Baltz

2016). A paradigm shift in the discovering process is helping scientists to overcome this impasse, while novel strategies such as genome mining tackle challenges such as activation and expression of silent or poorly expressed pathways (Iftime et al. 2016; Baltz 2017). This is why, along with diversity studies to build "the general inventory project," (see Chap. 1 in this book) we started planning screening strategies for hundreds of isolates obtained in a 3-year isolation effort.

That effort was guided to increase the probability to isolate rare actinomycetes producing interesting novel products. To achieve this goal, it has been recommended to design antibiotic-based culture media allowing taxon-selected isolation in under-explored environments (Baltz 2016). In a recent study in CCB, Arocha-Garza et al. (2017) used nine culture media designed to isolate actinobacteria. Sampling at the Churince system, a total of 105 morphologically diverse actinobacterial strains were successfully cultured. The isolates encompassed 11 genera: *Streptomyces*, *Arthrobacter*, *Cellulosimicrobium*, *Nocardia*, *Kocuria*, *Nocardiopsis*, *Microbacterium*, *Glycomyces*, *Blastococcus*, *Verrucosispora*, and *Micrococcus*. Among the recovered strains, *Streptomyces* was the most abundant genus forming 38 unique OTUs in 6 new monophyletic clades. These results gave our group the certainty of the uniqueness of the members of this phylum and provided us with the enthusiasm we needed to find financial resources to continue with the bioprospecting efforts.

The Tricks that CCB Microbes Can Do

It is not hard to forsee the wide diversity of applications for natural products that can potentially be obtained from Actinobacteria at CCB, from biomedical therapeutics and agro-industrial products, to bioremediation solutions. Our research group (Team Nuevo Leon) has focused on screening retrieved CCB microbes by setting up several preliminary assays that have helped us make a first scrutiny to filter hundreds of isolates and move along the screening pipeline to further evaluate and deeply characterize the most promising strains (Fig. 9.1).

Here we present these first results that have opened several possibilities in this bioprospecting story.

Antibiotics and Antifungals

We have seen for many years now the emergence of new drug-resistant pathogenic bacteria and fungi due to the indiscriminate use of antibiotics and antifungals in human therapy as well as in farm animals (Spizek et al. 2010), and the numbers are increasing constantly around the world. For 2004, it was estimated that more than 70% of pathogenic bacteria were resistant to at least one type of antibiotic (Demain and Sanchez 2009), and in February 2017, the World Health Organization published

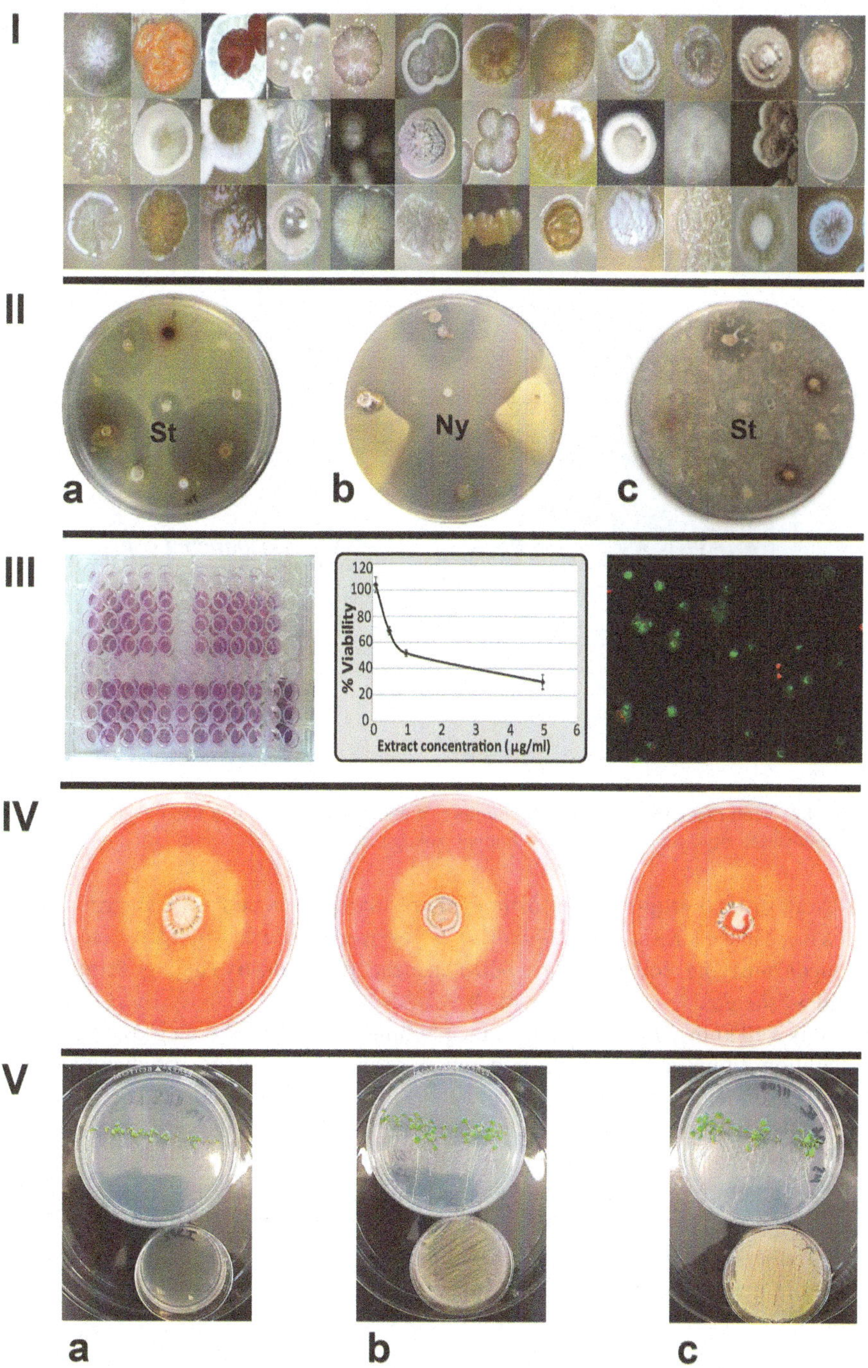

Fig. 9.1 Isolation and screening assays for the identification of CCB microbes with biotechnological potential. I. Actinobacteria retrieved from CCB ponds displaying diversity of morphologies. II. Antagonism assay of CCB actinobacteria against human pathogens (A. *Staphylococcus aureus*. B. *Candida albicans*. C. *Escherichia coli*). III. Antitumoral activity assay from crude extracts obtained from CCB microalgae and *Actinomycetes*. IV. Enzymatic hydrolysis of cellulose. V. Plant growth promoter (a. negative control. b. *Micrococcus* sp. c. *Kocuria* sp.)

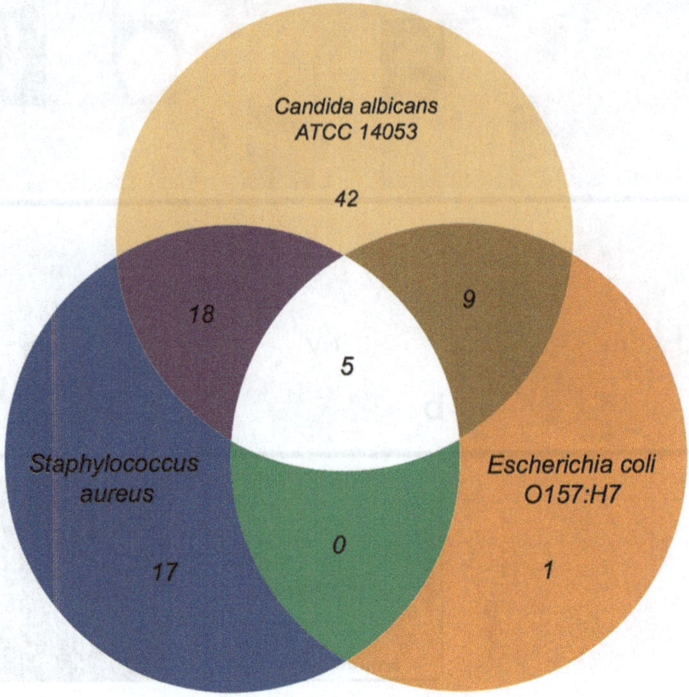

Fig. 9.2 Venn diagram that shows the number of actinobacteria isolated from CCB that exhibit inhibitory activity against one or more human pathogens

a list of critical drug-resistant pathogenic bacteria (Tacconelli and Magrini 2017); this has caused a resurgence of interest in discovering new biologically active compounds unifying collaborations between academy, industry, and government to achieve the goal.

It is well known that members of the phylum *Actinobacteria* can grow in extreme conditions of nutrient scarcity, temperature, salinity, and pressure, producing a wide number of metabolites for their survival that exhibit diverse biological activities. To avoid the rediscovery of known bioactive compounds, our isolation efforts have not been limited to one pond, and ongoing investigation has continued intensely designing new culture media and pretreatments, including variations in temperature, pH, antibiotics, and strong chemicals (Tiwari and Gupta 2013; Qin et al. 2009; Subramani and Aalbersberg 2013) to expand the number of actinobacterial strains of different and rare genera. So far, until 2018, the efforts have resulted in more than 500 isolates characterized as *Actinomycetes* (Fig. 9.1I).

Using Petri dishes and the *double-layer* agar *method* (Zheng et al. 2000) in four different culture media, we found that in this collection, 20% were antagonists against human pathogens such as *Candida albicans*, *Staphylococcus aureus*, and *Escherichia coli*. Up to 2017, we have found that 93 of the tested *Actinobacteria* inhibited the growth of at least 1 pathogen (Figs. 9.1II and 9.2).

We think that the fact that our strains have shown such a strong antagonistic behavior toward human pathogens explains why in none of our surveys for 17 years in CCB ponds, we have found any *E. coli* or *Salmonella* out of millions of sequences (V. Souza, personal communication) and the reason why the microbial communities are unique. Migration of unwanted foreigners is chastised with death (Souza et al. 2012, Chap. 6 in this book). It is also noticeable that actinomycetes from CCB are more antagonistic to fungi and gram-positive actinobacteria than to gram-negatives. These observations enforce the idea of niche splitting among more similar entities than between lineages that have a very different niche. Similar results have been observed for the *Bacillus* (Perez-Gutierrez et al. 2013). Anyhow, with the results of these screenings of CCB actinomycetes, we have opened the first door of biotechnology potential: novel sources of antibiotics.

Anticancer Potential

Cancer is one of the most serious human health problems. The current therapeutic methods for people with cancer are surgery, radiotherapy, immunotherapy, and chemotherapy (Baskar et al. 2012). Nevertheless, the high toxicity as well as all the undesirable side effects associated with cancer chemotherapy drugs urges further research to develop alternative therapies with fewer side effects (Housman et al. 2014; Sudha and Masilamani 2012).

Moreover, 42% of anticancer drugs used between 1981 and 2006 were obtained from natural products (Newman and Cragg 2007), and half of them are produced by actinobacteria (Busi and Pattnaik 2018). Besides, some of those metabolites, as the case of anthracyclines, have dual activity as antibiotic and antitumoral drugs (Hopp et al. 2008; Westman et al. 2012). For this reason, it was very tempting to also test the anticancer activity of the several bioactive CCB actinomycetes in our collection through tests of their organic crude extracts in antitumoral assays.

Up to date, some microalgal as well as 13 actinobacterial extracts (previously reported to show antimicrobial activity) have been tested in our lab for their antitumoral activity against a panel of human cancer cells. Two out of thirteen tested extracts showed significant activity against one or more cancer cell lines with a half minimal inhibitory concentration (IC_{50}) lower than the criteria of high cytotoxicity for crude extracts, as established by the US National Cancer Institute (NCI), which is an $IC_{50} < 30$ μg/mL (Sudha and Masilamani 2012). Meanwhile, we identified the strains that produced the bioactive metabolites as members of the genus *Streptomyces*. The 2 crude extracts selected by their high cytotoxicity activity were separated using flash chromatography, and the 12 components were tested again to find the responsible, active chemical fraction (Fig. 9.1III). Ongoing work in our group is trying to separate and identify the bioactive molecule, as well as finding in their genome the genes responsible for its synthesis. We opened a second important door for pharmaceutical development: natural compounds to battle cancer.

Enzymatic Hydrolysis of Cellulose

In the last decades, the use of energy has proven to be the major fraction of emission of greenhouse gases that is severely affecting our planet and menacing its future. On the one hand, lignocellulose constitutes more than 60 % of the plant biomass on Earth, of which the vast majority are agricultural residues that are confined without a use (Kumar et al. 2008). The compositions of these materials are different depending on the source. The main components are cellulose (35–50%), followed by hemicellulose (20–35%) and lignin (10–25%) (Xu and Wang 2015; Sun and Cheng 2002).

On the other hand, ethanol produced by lignocellulose materials from industrial crops or waste products could be a great solution to produce green energy. Nevertheless, cellulose, its principal component, is difficult to degrade (Palme et al. 2016; Pavasars et al. 2003). However, a few years ago, the use of cellulases produced by microorganisms seemed to be promising. Many microorganisms have been identified that are capable of hydrolyzing cellulose, but not all of them have the capacity to produce enough amounts of cellulases to scale production to the industry. Within the promising bacterial lineages, many genera from the phylum, such as *Streptomyces*, *Rhodococcus*, *Cellulosimicrobium*, *Actinoalloteichus*, and *Microbispora*, have been reported to produce a large amount of extracellular hydrolytic enzymes such endoglucanases, β-glucosidases, endoxylanases, and β-xylosidases (Liang et al. 2014; Gobalakrishnan and Sivakumar 2016; López-Mondéjar et al. 2016).

Being aware that actinobacterial strains have a huge potential to degrade cellulose, our team began to screen isolated actinobacteria from CCB in a first approach to identify the cellulolytic strains. Over 150 actinobacteria isolates were tested using a Congo red assay (Gobalakrishnan and Sivakumar 2016), and up to 31 isolates producing extracellular cellulolytic enzymes were identified, and 12 were considered for further analysis (Fig. 9.1IV), as in the near future, these could be used in the production of enzymes at an industrial level. A third door has been opened: solutions for sustainable energy production.

Plant Growth Promoters

Plant growth-promoting rhizobacteria (PGPR) is a group of bacteria that improve plant growth, increase yield, enhance soil fertility, and reduce pathogens as well as biotic or abiotic stresses (Bhattacharyya and Jha 2012; Saleem et al. 2007). Actinomycetes are present in the plant rhizosphere and produce various agro-active compounds such as indole acetic acid (IAA), cytokinins, gibberellins, siderophores, antimicrobials, and other compounds that participate in solubilization of inorganic phosphate and asymbiotic nitrogen fixation (Anwar et al. 2016). Actinobacteria, particularly the *Streptomyces* genus members, also exhibit immense biocontrol action against a wide range of phytopathogens (Wang et al. 2013) and are renowned for the production of an exceptionally large number of secondary metabolites that significantly enhance plant shoot and root biomass (Cordovez et al. 2015). However, in spite of actinobacterial high soil population, secondary metabolite production, and their ability to inhabit hostile environments, *Streptomyces* and other actinobacteria have been underexplored for plant growth promotion, compared to *Pseudomonas* and *Bacillus* species (Sousa and Olivares 2016).

We wanted to explore the potential role of our collection of *Actinomycetes* as plant growth promoters in *Arabidopsis thaliana*, a plant model widely used in the plant-rhizobacteria interaction (Desbrosses et al. 2009; van der Meij et al. 2018). Sterile seedlings in Petri dishes were exposed to CCB actinomycetes; results showed striking effects on plant development of some of the strains that are worth further evaluation, but, in brief, we have demonstrated that some of our strains significantly increase biomass of *A. thaliana* seedlings, whereas two isolates decreased biomass of the plants. Even if exploratory, these experiments unambiguously demonstrated the potential application of strains belonging to *Kocuria* and *Micrococcus* genera for sustainable agriculture and an alternative to bio-fertilizers (Fig. 9.1V) and strains of *Streptomyces* as potential biological herbicides. Currently, robust experiments are being carried out in our laboratory to determine whether those isolates are phosphorus solubilizers, phytohormone and siderophore producers, and phytopathogen controllers. A fourth door was opened with *Actinobacteria*: sustainable agriculture.

The Unexplored Path of Archaea and Halophiles

Archaea despite its important role in early Earth (see Chap. 2 in this book) are in general not very abundant in most ecosystems on Earth. Consistently, at CCB only four species of Archaea were found in the early samples, all undescribed (Souza et al. 2006). Three years later, a metagenomic analysis showed that the Archaea domain represents 1.8% of abundance in Pozas Rojas and 0.7% of abundance in Río Mezquites (Breitbart et al. 2009). Another metagenomic analysis of two microbial mats (described as the Red mat and the Green mat) showed that the abundance of Archaea domain in these samples was 0.26% and 2.0%, respectively (Peimbert et al. 2012). Microbial mats, as well as stromatolites, possess an undeniable interest in biotechnology due to their metabolic capacities (Kunin et al. 2008; Lau et al. 2009; Paerl et al. 2000; Peimbert et al. 2012). Currently, CCB diversity studies have reached 30 metagenomic studies, and all but one particular site show archaeal organisms as rare in the communities. This archaean-rich site in CCB has unique elastic microbial mats that we have called the "archaean domes," inside of which there is an anoxic atmosphere. The archaean time seems to be recreated in the domes, being dominated by methane along with sulfur and hydrocarbon volatiles (Chap. 2 this book).

Our group is working on isolation assays and metagenomic analysis as part of an ongoing project to elucidate how archaeal communities are different here from all the other sites at CCB and the world. Once those basic biodiversity questions are understood, the bioprospecting door for Archaea might be considered open.

The reason we had not spotted the "Archean domes" before is that there is a small window of few days a year, after heavy rain, during which these domes form, only in one small area of the whole Cuatro Cienegas Basin. Simply, that door had not been found, until it was opened in front of us in spring equinox 2016. The rest of the time in the year, the area is just a salty crust.

Halophiles, including Archaea and bacteria, are widely distributed in hypersaline environments, such as salt lakes, brine ponds, solar salterns, and salt mines, among others (Oren 2002). Just a few of halophile microorganisms have been isolated and studied (Meseguer-Soria 2004). The red-, orange-, or purple-pigmented bacteria or archaea could be either aerobic or facultative anaerobic and are potentially a source of high-value metabolites, thanks to their salt-tolerant enzymes, protective pigments, halocins, and even antitumor agents (De la Vega et al. 2016; Lanyi 2004; Schiraldi et al. 2002; Torreblanca et al. 1994). Currently, haloarchaea compounds are used in industry, for example, bacteriorhodopsin produced by *Halobacterium salinarum* has applications in holography, artificial retinas, and neural network optical computing, among others, due to its photoelectric, proton motive, and photochemical features. At present, two companies, COBEL (Barcelona) and Munich Innovative Biomaterials (MIB), sell purple membrane patches containing bacteriorhodopsin (Margesin and Schinner 2001); another commercial bacteriorhodopsin is sold as part of a motion bio/chemical sensor (Patent No. US5719033) (Ackley and Shieh 1998). Patented *liposomes* (No. US5989587), from *Halobacterium cutirubrum*, are used in the medical field as transporters of drugs to target specific sites in the body (Sprott et al. 1999). And, finally, enzymes. These halophilic enzymes include isomerases such as cyclophilin-type PPIase, currently in the market (Iida et al. 1997), and hydrolases (α-amylases, proteases, β-galactosidases) (Margesin and Schinner 2001). Even if the Archaea domain does not possess NRPS gene clusters for biosynthesis of secondary metabolites, they can display many survival mechanism for the synthesis of natural products (Charlesworth and Burns 2015) that are worth exploring.

Moreover, there are many commercial extremozymes (enzymes from extremophiles) available. Not only are halophiles a source of potential novel enzymes; psychrophiles and thermophiles do it as well. These enzymes can display a better performance in standard tough industrial processes because they do not denature easily. For example, cold-active enzymes used in the industrial market are Lipoclean® and Stainzyme® Plus (Novozymes), Excellase® and Purefect® OxAm (Genecor), and Effectenz® and Preferenz® (DuPont) all applied in the detergents' market. Also, commercial thermophilic enzymes have presence in industrial applications, for their efficiency under high temperatures such as Avantec®, Spirizyme®, and GluZym® (Novozymes), used in food and beverage market, and Lacasse® and Lipase B Lypozyme® (Novozymes) applied in pulp and paper production (Sarmiento et al. 2015).

An ongoing collaborative effort of haloarchaea isolation from CCB is on its way. So far, we have successfully isolated and characterized an haloarchaea belonging to the genus *Halorobrum*, according to 16S rRNA results. The aforementioned strain grows optimally with NaCl concentrations of 20–25%, both on solid (agar) and liquid media, in about 4 weeks; the macroscopic morphology of the strain is a circular and convex colony with a dual pigmentation (yellow and red), and microscopically, they are gram-negative and catalase- and oxidase-positive. This extremely halophilic archaeon could have common industrial applications such as food additives and microencapsulation (lycopene); as diet supplements and coloring (β-carotene); in starch bioprocessing, like laundry detergent additives; and in the baking industry (pullulanase), but also in synthetic chemistry and ethanol production (β-galactosidase) (Mohammadipanah and Wink 2015; Oren et al. 2018; Olsson et al. 2017; Peck et al. 2017; Laye and DasSarma 2017).

Perspectives

It is hard to give closure to this chapter since it is barely an opening to what we expect, will become a start point to several successful stories for microorganisms isolated from CCB. These unique strains will open new questions to gain biological knowledge and will provide with solutions to human problems through science. Not only have we been able to successfully culture a significant number of unique actinomycetes from a single pond from CCB, but there are still important biotechnological areas that have not been explored. Moreover, there are still other strategies that must be applied to successfully scrutinize is huge microbial diversity in order to be able to uncover the amazing potential waiting to be discovered. It will be necessary to invest in concerted long-term efforts to search for genuine and substantial contributions that will let us move this recently discovered potential toward truly transferred biotechnology. By following the Nagoya treaty, we can envision a "biotech valley" where the local kids participate in making the discoveries, creating value for their biodiversity, and returning to their community the necessary resources for education and sustainable development to alleviate poverty. The only way that such a future is possible is by protecting the wetland and its precious hidden treasure.

Acknowledgments This chapter was written during a sabbatical leave of VS in the Department of Ecology, Evolution, and Behavior, University of Minnesota, in Dr. Michael Travisano's laboratory supported by the program PASPA-DGAPA, UNAM.

References

Ackley DE, Shieh C-L (1998) Thin film transistor bio/chemical sensor. Motorola Solutions Inc, Schaumburg

Alcaraz LD, Olmedo G, Bonilla G, Cerritos R, Hernandez G, Cruz A, Ramirez E et al (2008) The genome of Bacillus coahuilensis reveals adaptations essential for survival in the relic of an ancient marine environment. Proc Natl Acad Sci U S A 105(15):5803–5808. https://doi.org/10.1073/pnas.0800981105

Anwar S, Ali B, Sajid I (2016) Screening of rhizospheric actinomycetes for various in-vitro and in-vivo plant growth promoting (PGP) traits and for agroactive compounds. Front Microbiol 7:1334. https://doi.org/10.3389/fmicb.2016.01334

Arocha-Garza HF, Canales-Del Castillo R, Eguiarte LE, Souza V, De la Torre-Zavala S (2017) High diversity and suggested endemicity of culturable actinobacteria in an extremely oligotrophic desert oasis. PeerJ 5:e3247. https://doi.org/10.7717/peerj.3247

Avitia M, Escalante AE, Rebollar EA, Moreno-Letelier A, Eguiarte LE, Souza V (2014) Population expansions shared among coexisting bacterial lineages are revealed by genetic evidence. PeerJ 2:e696. https://doi.org/10.7717/peerj.696

Baltz RH (2016) Gifted microbes for genome mining and natural product discovery. J Ind Microbiol Biotechnol 44(4-5):573–588. https://doi.org/10.1007/s10295-016-1815-x

Baltz RH (2017) Synthetic biology, genome mining, and combinatorial biosynthesis of NRPS-derived antibiotics: a perspective. J Ind Microbiol Biotechnol. https://doi.org/10.1007/s10295-017-1999-8

Baskar R, Lee KA, Yeo R, Yeoh KW (2012) Cancer and radiation therapy: current advances and future directions. Int J Med Sci 9(3):193–199. https://doi.org/10.7150/ijms.3635

Bhattacharyya PN, Jha DK (2012) Plant growth-promoting rhizobacteria (PGPR): emergence in agriculture. World J Microbiol Biotechnol 28(4):1327–1350. https://doi.org/10.1007/s11274-011-0979-9

Breitbart M, Hoare A, Nitti A, Siefert J, Haynes M, Dinsdale E, Edwards R, Souza V, Rohwer F, Hollander D (2009) Metagenomic and stable isotopic analyses of modern freshwater microbialites in Cuatro Cienegas, Mexico. Environ Microbiol 11(1):16–34. https://doi.org/10.1111/j.1462-2920.2008.01725.x

Busi S, Pattnaik S (2018) Current status and applications of actinobacteria in the production of anticancerous compounds. In: Actinobacteria: diversity and biotechnological applications. Elsevier, Oxford

Cerritos R, Eguiarte LE, Avitia M, Siefert J, Travisano M, Rodriguez-Verdugo A, Souza V (2011) Diversity of culturable thermo-resistant aquatic bacteria along an environmental gradient in Cuatro Cienegas, Coahuila, Mexico. Antonie Van Leeuwenhoek 99(2):303–318. https://doi.org/10.1007/s10482-010-9490-9

Charlesworth JC, Burns BP (2015) Untapped resources: biotechnological potential of peptides and secondary metabolites in Archaea. Archaea 2015:282035. https://doi.org/10.1155/2015/282035

Cordovez V, Carrion VJ, Etalo DW, Mumm R, Zhu H, van Wezel GP, Raaijmakers JM (2015) Diversity and functions of volatile organic compounds produced by Streptomyces from a disease-suppressive soil. Front Microbiol 6:1081. https://doi.org/10.3389/fmicb.2015.01081

Cruz-Morales P, Ramos-Aboites HE, Licona-Cassani C, Selem-Mojica N, Mejia-Ponce PM, Souza-Saldivar V, Barona-Gomez F (2017) Actinobacteria phylogenomics, selective isolation from an iron oligotrophic environment and siderophore functional characterization, unveil new desferrioxamine traits. FEMS Microbiol Ecol 93(9):fix086. https://doi.org/10.1093/femsec/fix086

De la Vega M, Sayago A, Ariza J, Barneto AG, Leon R (2016) Characterization of a bacterioruberin-producing Haloarchaea isolated from the marshlands of the Odiel river in the southwest of Spain. Biotechnol Prog 32(3):592–600. https://doi.org/10.1002/btpr.2248

Demain AL, Sanchez S (2009) Microbial drug discovery: 80 years of progress. J Antibiot 62(1):5–16. https://doi.org/10.1038/ja.2008.16

Desbrosses G, Contesto C, Varoquaux F, Galland M, Touraine B (2009) PGPR-Arabidopsis interactions is a useful system to study signaling pathways involved in plant developmental control. Plant Signal Behav 4(4):321–323

Dominguez-Escobar J, Beltran Y, Bergman B, Diez B, Ininbergs K, Souza V, Falcon LI (2011) Phylogenetic and molecular clock inferences of cyanobacterial strains within Rivulariaceae from distant environments. FEMS Microbiol Lett 316(2):90–99. https://doi.org/10.1111/j.1574-6968.2010.02195.x

Escalante AE, Caballero-Mellado J, Martinez-Aguilar L, Rodriguez-Verdugo A, Gonzalez-Gonzalez A, Toribio-Jimenez J, Souza V (2009) Pseudomonas cuatrocienegasensis sp. nov., isolated from an evaporating lagoon in the Cuatro Cienegas valley in Coahuila, Mexico. Int J Syst Evol Microbiol 59(Pt 6):1416–1420. https://doi.org/10.1099/ijs.0.006189-0.

Gobalakrishnan R, Sivakumar K (2016) Systematic characterization of potential cellulolytic marine actinobacteria Actinoalloteichus sp. MHA15. Biotechnol Rep 13:30–36

Hopp DC, Rabenstein J, Rhea J, Smith C, Romari K, Clarke M, Francis L et al (2008) Mutactimycin E, a new anthracycline antibiotic with gram-positive activity. J Antibiot 61(11):675–679. https://doi.org/10.1038/ja.2008.95

Housman G, Byler S, Heerboth S, Lapinska K, Longacre M, Snyder N, Sarkar S (2014) Drug resistance in cancer: an overview. Cancer 6(3):1769–1792. https://doi.org/10.3390/cancers6031769

Iftime D, Kulik A, Hartner T, Rohrer S, Niedermeyer TH, Stegmann E, Weber T, Wohlleben W (2016) Identification and activation of novel biosynthetic gene clusters by genome mining in the kirromycin producer Streptomyces collinus Tu 365. J Ind Microbiol Biotechnol 43(2-3):277–291. https://doi.org/10.1007/s10295-015-1685-7

Iida T, Furyiya M, Suzuki K, Iwabuchi N, Maruyama T (1997) Cyclophilin type PPI-ase gene originating from halophilic archaebacterium. Japan

Katz L, Baltz RH (2016) Natural product discovery: past, present, and future. J Ind Microbiol Biotechnol 43(2-3):155–176. https://doi.org/10.1007/s10295-015-1723-5

Kumar R, Singh S, Singh OV (2008) Bioconversion of lignocellulosic biomass: biochemical and molecular perspectives. J Ind Microbiol Biotechnol 35(5):377–391. https://doi.org/10.1007/s10295-008-0327-8

Kunin V, Raes J, Harris JK, Spear JR, Walker JJ, Ivanova N, von Mering C et al (2008) Millimeter-scale genetic gradients and community-level molecular convergence in a hypersaline microbial mat. Mol Syst Biol 4:198. https://doi.org/10.1038/msb.2008.35

Lanyi JK (2004) Bacteriorhodopsin. Annu Rev Physiol 66:665–688. https://doi.org/10.1146/annurev.physiol.66.032102.150049

Lau MC, Aitchison JC, Pointing SB (2009) Bacterial community composition in thermophilic microbial mats from five hot springs in central Tibet. Extremophiles 13(1):139–149. https://doi.org/10.1007/s00792-008-0205-3

Laye VJ, DasSarma S (2017) An antarctic extreme halophile and its polyextremophilic enzyme: effects of perchlorate salts. Astrobiology 18(4):412–418. https://doi.org/10.1089/ast.2017.1766

Lee ZM, Steger L, Corman JR, Neveu M, Poret-Peterson AT, Souza V, Elser JJ (2015) Response of a stoichiometrically imbalanced ecosystem to manipulation of nutrient supplies and ratios. PLoS One 10(4):e0123949. https://doi.org/10.1371/journal.pone.0123949

Liang Y-L, Zhang Z, Wu M, Wu Y, Feng J-X (2014) Isolation, screening, and identification of cellulolytic bacteria from natural reserves in the subtropical region of China and optimization of cellulase production by Paenibacillus terrae ME27-1. Hindawi 2014:512497

López-Mondéjar R, Zühlke D, Becher D, Riedel K, Baldrian P (2016) Cellulose and hemicellulose decomposition by forest soil bacteria proceeds by the action of structurally variable enzymatic systems. Sci Rep 6:25279

Margesin R, Schinner F (2001) Potential of halotolerant and halophilic microorganisms for biotechnology. Extremophiles 5(2):73–83

Meseguer-Soria I (2004) Los microorganismos halófilos y su potencial aplicado en biotecnología. Cienc Invest 7(2):13–17

Mohammadipanah F, Wink J (2015) Actinobacteria from arid and desert habitats: diversity and biological activity. Front Microbiol 6:1541. https://doi.org/10.3389/fmicb.2015.01541.

Moreno-Letelier A, Olmedo-Alvarez G, Eguiarte LE, Souza V (2012) Divergence and phylogeny of Firmicutes from the Cuatro Cienegas Basin, Mexico: a window to an ancient ocean. Astrobiology 12(7):674–684. https://doi.org/10.1089/ast.2011.0685

Newman DJ, Cragg GM (2007) Natural products as sources of new drugs over the last 25 years. J Nat Prod 70(3):461–477. https://doi.org/10.1021/np068054v

Olsson BE, Korsakova ES, Anan'ina LN, Pyankova AA, Mavrodi OV, Plotnikova EG, Mavrodi DV (2017) Draft genome sequences of strains Salinicola socius SMB35(T), Salinicola sp.

MH3R3-1 and Chromohalobacter sp. SMB17 from the Verkhnekamsk potash mining region of Russia. Stand Genomic Sci 12:39. https://doi.org/10.1186/s40793-017-0251-5

Oren A, Hirschberg J, Mann V, Jehlicka J (2018) Effects of nicotine on the biosynthesis of carotenoids in halophilic Archaea (class Halobacteria): an HPLC and Raman spectroscopy study. Extremophiles 22(3):359–366. https://doi.org/10.1007/s00792-018-0995-x

Oren A (2002) Halophilic microorganisms and their environments. 1 Aufl. Cellular origin, life in extreme habitats and astrobiology, vol 5. Kluwer Academic Publishers, Boston

Paerl HW, Pinckney JL, Steppe TF (2000) Cyanobacterial-bacterial mat consortia: examining the functional unit of microbial survival and growth in extreme environments. Environ Microbiol 2(1):11–26

Palme A, Theliander H, Brelid H (2016) Acid hydrolysis of cellulosic fibres: comparison of bleached kraft pulp, dissolving pulps and cotton textile cellulose. Carbohydr Polym 136:1281–1287. https://doi.org/10.1016/j.carbpol.2015.10.015

Pavasars I, Hagberg J, Borén H, Allard B (2003) Alkaline degradation of cellulose: mechanisms and kinetics. Kluwer Academic Publishers, Dordrecht

Peck RF, Plesa AM, Graham SM, Angelini DR, Shaw EL (2017) Opsin-mediated inhibition of bacterioruberin synthesis in halophilic Archaea. J Bacteriol 199(21):e00303–e00317. https://doi.org/10.1128/JB.00303-17

Peimbert M, Alcaraz LD, Bonilla-Rosso G, Olmedo-Alvarez G, Garcia-Oliva F, Segovia L, Eguiarte LE, Souza V (2012) Comparative metagenomics of two microbial mats at Cuatro Cienegas Basin I: ancient lessons on how to cope with an environment under severe nutrient stress. Astrobiology 12(7):648–658. https://doi.org/10.1089/ast.2011.0694

Perez-Gutierrez RA, Lopez-Ramirez V, Islas A, Alcaraz LD, Hernandez-Gonzalez I, Olivera BC, Santillan M et al (2013) Antagonism influences assembly of a Bacillus guild in a local community and is depicted as a food-chain network. ISME J 7(3):487–497. https://doi.org/10.1038/ismej.2012.119

Qin S, Li J, Chen HH, Zhao GZ, Zhu WY, Jiang CL, Xu LH, Li WJ (2009) Isolation, diversity, and antimicrobial activity of rare actinobacteria from medicinal plants of tropical rain forests in Xishuangbanna, China. Appl Environ Microbiol 75(19):6176–6186. https://doi.org/10.1128/AEM.01034-09

Rebollar EA, Avitia M, Eguiarte LE, Gonzalez-Gonzalez A, Mora L, Bonilla-Rosso G, Souza V (2012) Water-sediment niche differentiation in ancient marine lineages of Exiguobacterium endemic to the Cuatro Cienegas Basin. Environ Microbiol 14(9):2323–2333. https://doi.org/10.1111/j.1462-2920.2012.02784.x

Saleem M, Arshad M, Hussain S, Bhatti AS (2007) Perspective of plant growth promoting rhizobacteria (PGPR) containing ACC deaminase in stress agriculture. J Ind Microbiol Biotechnol 34(10):635–648. https://doi.org/10.1007/s10295-007-0240-6

Sarmiento F, Peralta R, Blamey JM (2015) Cold and hot extremozymes: industrial relevance and current trends. Front Bioeng Biotechnol 3:148. https://doi.org/10.3389/fbioe.2015.00148

Schiraldi C, Giuliano M, De Rosa M (2002) Perspectives on biotechnological applications of archaea. Archaea 1(2):75–86

Sousa JA d J, Olivares FL (2016) Plant growth promotion by streptomycetes: ecophysiology, mechanisms and applications. Chem Biol Technol Agric 3(1):24. https://doi.org/10.1186/s40538-016-0073-5

Souza V, Eguiarte LE, Travisano M, Elser JJ, Rooks C, Siefert JL (2012) Travel, sex, and food: what's speciation got to do with it? Astrobiology 12(7):634–640. https://doi.org/10.1089/ast.2011.0768

Souza V, Espinosa-Asuar L, Escalante AE, Eguiarte LE, Farmer J, Forney L, Lloret L et al (2006) An endangered oasis of aquatic microbial biodiversity in the Chihuahuan desert. Proc Natl Acad Sci U S A 103(17):6565–6570. https://doi.org/10.1073/pnas.0601434103

Spizek J, Novotna J, Rezanka T, Demain AL (2010) Do we need new antibiotics? The search for new targets and new compounds. J Ind Microbiol Biotechnol 37(12):1241–1248. https://doi.org/10.1007/s10295-010-0849-8

Sprott DG, Patel GB, Choquet CG, Ekiel I (1999) Formation of stable liposomes from lipid extracts of archaeobacteria. National Research Council of Canada. United States

Subramani R, Aalbersberg W (2013) Culturable rare actinomycetes: diversity, isolation and marine natural product discovery. Appl Microbiol Biotechnol 97(21):9291–9321. https://doi.org/10.1007/s00253-013-5229-7

Sudha S, Masilamani SM (2012) Characterization of cytotoxic compound from marine sediment derived actinomycete Streptomyces avidinii strain SU4. Asian Pac J Trop Biomed 2(10):770–773. https://doi.org/10.1016/S2221-1691(12)60227-5

Sun Y, Cheng J (2002) Hydrolysis of lignocellulosic materials for ethanol production: a review. Bioresour Technol 83(1):1–11

Tacconelli E, Magrini N (2017) In: Cavaleri M (ed) Global priority list of antibiotic-resistant bacteria to guide research, discovery, and development of new antibiotics. World Health Organization, Geneva

Tiwari K, Gupta RK (2012) Rare actinomycetes: a potential storehouse for novel antibiotics. Crit Rev Biotechnol 32(2):108–132. https://doi.org/10.3109/07388551.2011.562482

Tiwari K, Gupta RK (2013) Diversity and isolation of rare actinomycetes: an overview. Crit Rev Microbiol 39(3):256–294. https://doi.org/10.3109/1040841X.2012.709819

Toribio J, Escalante AE, Caballero-Mellado J, Gonzalez-Gonzalez A, Zavala S, Souza V, Soberon-Chavez G (2011) Characterization of a novel biosurfactant producing Pseudomonas koreensis lineage that is endemic to Cuatro Cienegas Basin. Syst Appl Microbiol 34(7):531–535. https://doi.org/10.1016/j.syapm.2011.01.007

Torreblanca M, Meseguer I, Ventosa A (1994) Production of halocin is a practically universal feature of archaeal halophilic rods. Lett Appl Microbiol 19(4):201–205. https://doi.org/10.1111/j.1472-765X.1994.tb00943.x

van der Meij A, Willemse J, Schneijderberg MA, Geurts R, Raaijmakers JM, van Wezel GP (2018) Inter- and intracellular colonization of Arabidopsis roots by endophytic actinobacteria and the impact of plant hormones on their antimicrobial activity. Antonie Van Leeuwenhoek 111(5):679–690. https://doi.org/10.1007/s10482-018-1014-z

Wang XJ, Zhang J, Wang JD, Qian PT, Liu CX, Xiang WS (2013) Novel cyclopentenone derivatives produced by a rare actinobacterial strain Actinoalloteichus nanshanensis sp. nov. NEAU 119. Nat Prod Res 27(20):1863–1869. https://doi.org/10.1080/14786419.2013.771349

Westman EL, Canova MJ, Radhi IJ, Koteva K, Kireeva I, Waglechner N, Wright GD (2012) Bacterial inactivation of the anticancer drug doxorubicin. Chem Biol 19(10):1255–1264. https://doi.org/10.1016/j.chembiol.2012.08.011

Xu F, Wang D (2015) Chapter 2. Analysis of lignocellulosic biomass using infrared methodology. In: Pretreatment of biomass. Elsevier, New York

Zheng Z, Zeng W, Huang Y, Yang Z, Li J, Cai H, Su W (2000) Detection of antitumor and antimicrobial activities in marine organism associated actinomycetes isolated from the Taiwan Strait, China. FEMS Microbiol Lett 188(1):87–91

Index

V. Souza et al. (eds.), *Cuatro Ciénegas Ecology, Natural History and Microbiology*, Cuatro Ciénegas Basin: An Endangered Hyperdiverse Oasis,
https://doi.org/10.1007/978-3-319-93423-5

N

O

P

W

X

Y

The manufacturer's authorised representative in the EU is Springer Nature Customer Service Centre GmbH, Europaplatz 3, 69115 Heidelberg, Germany. If you have any concerns regarding our products, please contact ProductSafety@springernature.com

Printed and bound by CPI Group (UK) Ltd, Croydon, CR0 4YY
25/09/2026
02222855-0017